Foundations of Organic Chemistry

Michael Hornby

Stowe School, Buckingham

Josephine Peach

Somerville College and the Dyson Perrins Laboratory, University of Oxford

Series sponsor: **ZENECA**

ZENECA is a major international company active in four main areas of business: Pharmaceuticals, Agrochemicals and Seeds, Specialty Chemicals, and Biological Products.

ZENECA's skill and innovative ideas in organic chemistry and bioscience create products and services which improve the world's health, nutrition, environment, and quality of life.

ZENECA is committed to the support of education in chemistry and chemical engineering.

OXFORD
UNIVERSITY PRESS

Great Clarendon Street, Oxford OX2 6DP
Oxford University Press is a department of the University of Oxford.
It furthers the University's objective of excellence in research, scholarship, and education by publishing worldwide in

Oxford New York
Athens Auckland Bangkok Bogotá Buenos Aires Calcutta
Cape Town Chennai Dar es Salaam Delhi Florence Hong Kong Istanbul
Karachi Kuala Lumpur Madrid Melbourne Mexico City Mumbai
Nairobi Paris São Paolo Singapore Taipei Tokyo Toronto Warsaw

with associated companies in Berlin Ibadan

Oxford is a registered trade mark of Oxford University Press
in the UK and in certain other countries

Published in the United States
by Oxford University Press Inc., New York

First published 1993
Reprinted 1994 (twice, with corrections), 1995 (with corrections), 1996, 1997, 1999, 2000

A catalogue record for this book is available from the British Library

Library of Congress Cataloging in Publication Data
Hornby, G. Michael
Foundations of organic chemistry / G. Michael Hornby and Josephine M. Peach
(Oxford chemistry primers: 9)
1. Chemistry, Organic. I. Peach, Josephine M. II. Title.
III. Series.
QD253.H67 1993 547—dc20 92-26902

ISBN 0 19 855680 2 (Pbk)

Printed in Great Britain by
The Bath Press, Bath

Series Editor's Foreword

Students starting chemistry courses at university have a wide range of backgrounds, and consequently each individual's elementary knowledge of the subject tends to be fragmented differently. Ideally all chemistry students should start their university courses with the same elementary knowledge base. This Oxford Chemistry Primer is designed to draw the subject together to provide a concise introduction to organic chemistry. It should stimulate young people reading chemistry at advanced level in school, and serve to excite still further the student's interest in the subject during the transitional period between school and university.

This primer will be of interest to chemistry schoolteachers and all who aspire to serve an apprenticeship in chemistry at university.

Stephen G. Davies
The Dyson Perrins Laboratory, University of Oxford

Preface

This book has been written to bridge the gap between organic chemistry at school and at university, to stimulate the interest of advanced level chemists, and to provide a foundation for undergraduates starting courses in chemistry or biochemistry. The first three chapters lay the basis of the physical chemistry needed and lead into the discussion of the reactions from a mechanistic point of view. We have tried to keep the majority of the main text within the A level syllabuses, with some extensions in the margins and in the last chapter. It is not intended as, nor can it be, a comprehensive A level text. By concentrating on mechanism we hope to emphasize the common threads that hold the subject together.

We are most grateful to all those who have given us valuable criticism and advice, in particular Dr Peter Carpenter (Roedean School), Dr John Nixon (Haberdashers' Aske's School, Elstree), and Dr David Smith (Winchester College). Dr Sydney Bailey (St Peter's College, Oxford) has been a constant source of help and inspiration to both of us since our own undergraduate days. Colleagues at ICI Pharmaceuticals and Agrochemicals have generously provided information on some of the commercially important compounds. Finally we would like to thank Mrs Brenda Armstrong (Fellows' Secretary at Somerville College, Oxford) for her valiant efforts at deciphering not one but two sets of chemical handwriting.

We would like to dedicate this book to Janet and to John, and to our mutual goddaughters, Helen and Emma.

Buckingham G. M. H.
Oxford J. M. P.
September 1992

Contents

1 Molecules 1

2 Mechanisms 18

3 Acids and bases 33

4 Reactions with nucleophiles 40

5 Reactions with electrophiles 58

6 Reactions with radical intermediates 75

7 Taking it further 83

Further reading 91

Index 91

1 Molecules

1.1 Introduction

Organic chemistry began as the chemistry of living things, like yeast, moulds, mushrooms, willow trees, and whales. It is based on the compounds of carbon, whose molecules are the essence of organic chemistry. From fuels through pain-killers and antibiotics to vitamins, they fascinate us. How are they constructed? How can we make them? Why do they behave the way they do? What can we make from them? How do we make molecules which are going to affect biological systems in the way we want, with minimal side-effects?

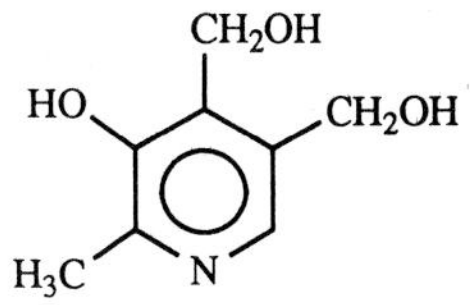

Vitamin B6

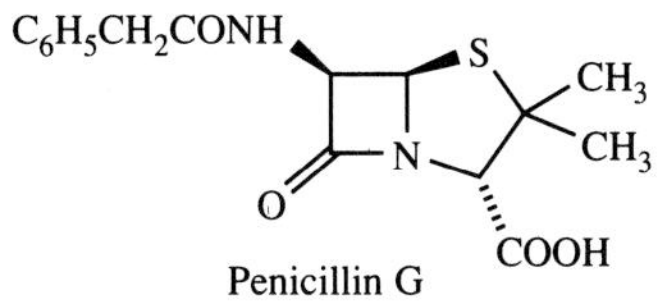

Penicillin G

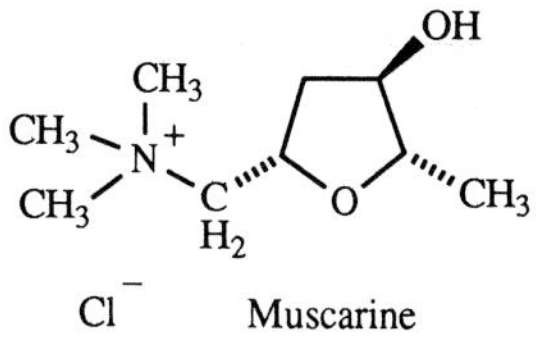

Muscarine

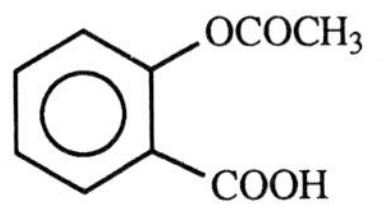

Aspirin

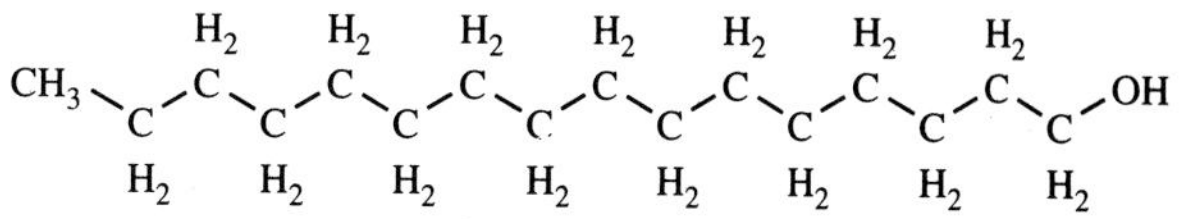

Cetyl alcohol

Before we can begin to answer these questions we need to understand the ways in which atoms join together to form molecules. In later chapters we will explore the ways in which a molecule's architecture affects its chemistry, and how we exploit this to make new compounds.

The abbreviated way of drawing organic molecules is explained on pp. 7 and 11.

1.2 Atoms

Chemical reactions are the result of bond breaking and bond making, involving the most energetic, outer shell electrons of atoms. Before we can discuss bonding we need to have some idea of how electrons are distributed in separate atoms, particularly carbon atoms.

The electronic configuration of a carbon atom in its ground state, or lowest energy state, is $1s^22s^22p^2$. The terms s and p describe the types of orbital in which these electrons are found, an orbital simply showing the probable distribution of electron density.

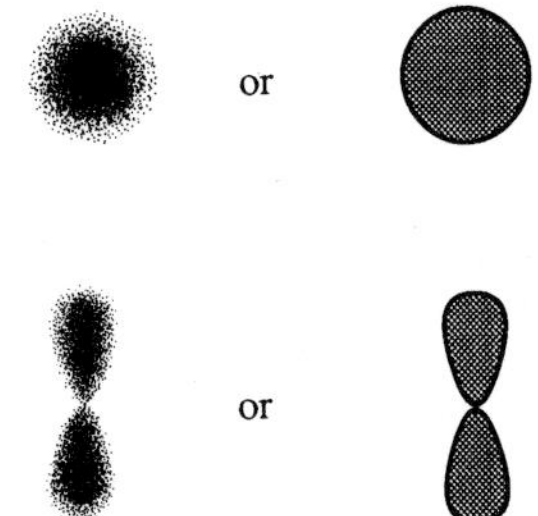

s Atomic orbitals

Each s orbital can hold a pair of electrons and is spherically symmetrical. The diagram shows the region in which an s electron is most likely to be found.

p Atomic orbitals

In the same way the electron density in a p orbital is shown as two pears end to end. p Orbitals are *not* spherically symmetrical and there are three of them, of equal energy, mutually at right angles, in each main shell.

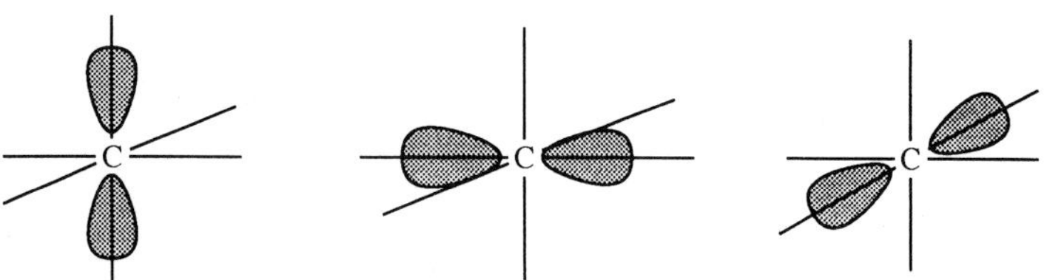

The electrons in a p orbital are most likely to be found somewhere within the 'double pear' region. On average they will be further from the nucleus than electrons in the corresponding s orbital.

1.3 Bonding

σ Bonds

The simplest case is hydrogen, H_2. The two singly occupied s orbitals on two hydrogen atoms combine to form one bonding orbital in H_2.

Two atomic orbitals combine to form *two* molecular orbitals, but only one is occupied and bonding.

Bonds like these, where two half-filled orbitals on neighbouring atoms combine to give a bonding molecular orbital with high electron density *between* the two nuclei involved, are called sigma or σ bonds. This is the normal single bond.

Making four single bonds to carbon

If a carbon atom remained in its ground state in a molecule it could form only two covalent bonds, using its two unpaired p electrons, and it would still have only six electrons in its outer shell.

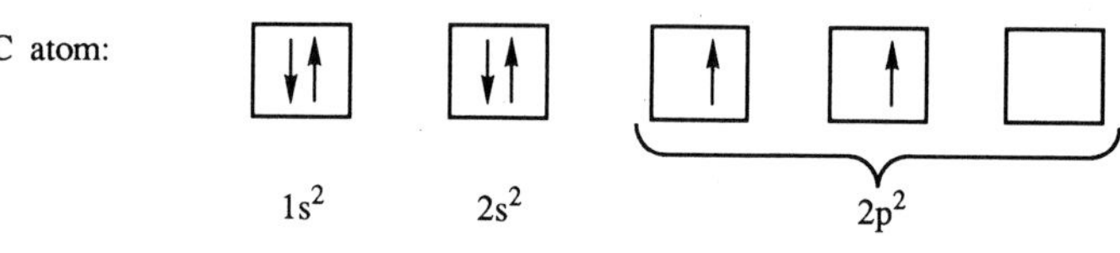

It is energetically more favourable to promote one of the 2s electrons into the vacant p orbital. The four half-filled orbitals can then combine with half-filled orbitals on four other atoms (e.g. H) to form *four equivalent covalent bonds*. The carbon atom makes four bonds and so will share eight electrons in its outer shell (two electrons per bond).

Again the eight atomic orbitals combine to form eight molecular orbitals, of which four are occupied and bonding.

In general *n* atomic orbitals combine to form *n* molecular orbitals, of which *n*/2 are occupied and bonding.

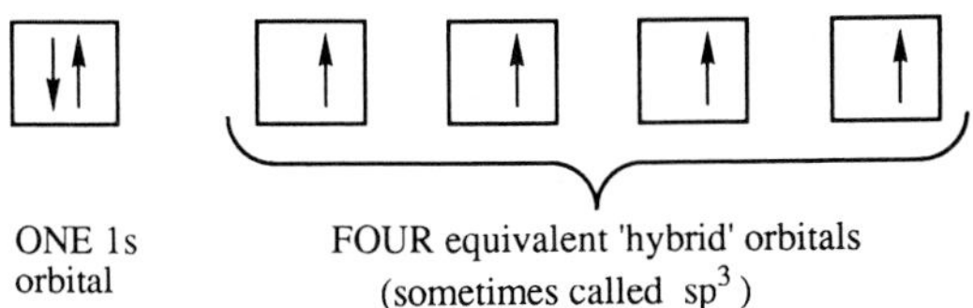

The energy for the promotion, about +400 kJ mol^{-1}, is more than repaid by the bond energy (see p. 10) from making four bonds instead of two.

Methane. In methane, CH_4, the four equivalent half-filled orbitals on carbon overlap with the half-filled s orbitals of four hydrogen atoms to give four new bonding molecular orbitals, each containing a shared pair of electrons. We have made four new σ bonds (see right).

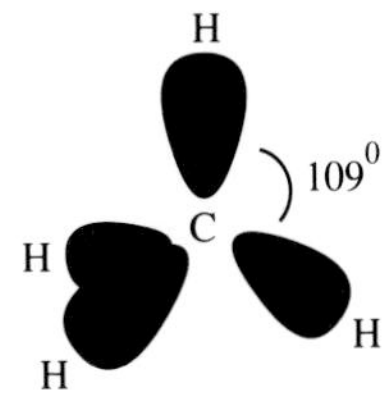

The shape of methane. Mutual repulsion of the electron pairs in these molecular orbitals or bonds leads to their pointing to the corners of a tetrahedron. This is the geometry for maximum separation of electrons and maximum bond angles (see p. 11).

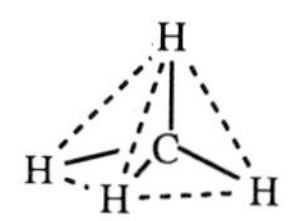

Making multiple bonds to carbon: π bonds

There are other ways in which the four half-filled orbitals of carbon can be combined, allowing carbon to make double or triple bonds to another atom.

Double bonds. Three of the four half-filled orbitals are used to form *three* equivalent 'hybrid' orbitals (called sp^2) leaving *one* half-filled p orbital to form one other bond.

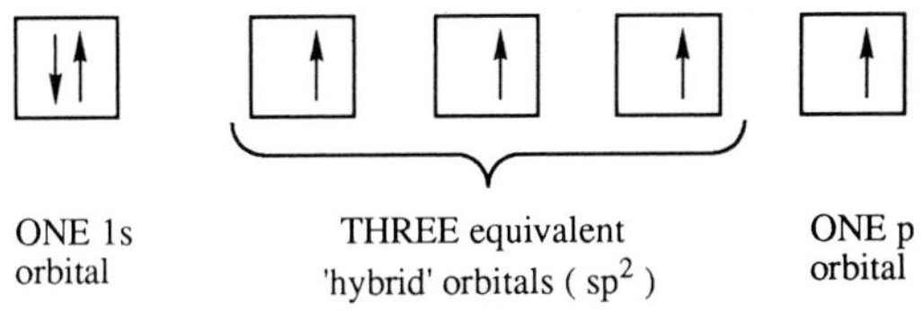

In ethene, C_2H_4, each carbon atom makes three σ bonds, using the three (sp^2) 'hybrid' orbitals.

sigma bonds: H, H, C, C or H, H, C–C

If the remaining half-filled p orbitals on the two carbon atoms are parallel, they can combine to form a new bonding orbital: a π (pi) bond. The shared electron pair is most likely to be found in the two sausage-shaped lobes on either side of the line joining the two nuclei.

π Bonds are always associated with a σ bond directly between the nuclei, giving a 'hamburger' look to the electron density diagram.

Together the σ and π bonds make a double bond between the two carbon atoms.

The shape of ethene. The principle of mutual repulsion of electron pairs helps to explain why the three atoms bonded to each carbon are in a trigonal planar arrangement (flat, propeller-like), with bond angles of about 120°. Ethene, C_2H_4, is a flat or planar molecule.

This diagram shows each shared pair of electrons as a line but makes no distinction between the σ and π contributions to the double bond, which is fundamental to an understanding of the reactions of ethene (see Chapter 5, p. 58). We will consider the implications of π bonding on the shapes of molecules later in this chapter (p. 12). Similar π bonds occur in the carbonyl group, C=O, in aldehydes, ketones, esters, and so on. The geometry is again planar with bond angles of about 120° (see left).

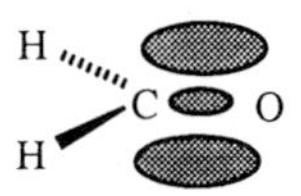

Triple bonds. Two of the half-filled orbitals on carbon form *two* equivalent 'hybrid' orbitals (called sp) leaving *two* half-filled p orbitals to form two π bonds.

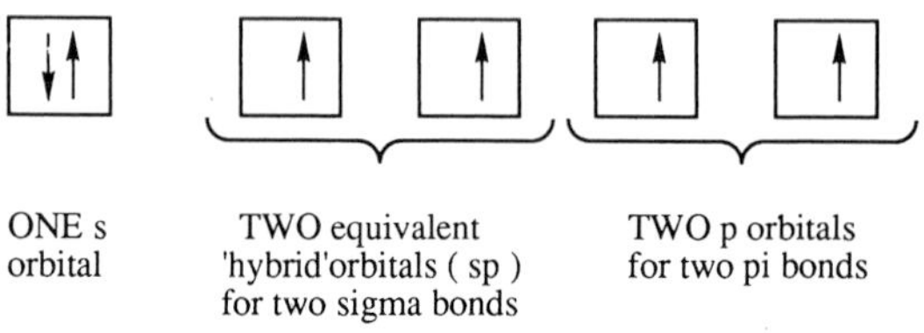

Linear triple bonds in alkynes and nitriles are the result of the formation of two π bonds using *two* mutually perpendicular p orbitals on each of the triple bonded atoms, with a σ bond in the middle as before. So the two π bonds of the triple bond are in planes at right angles to each other.

The distribution of the π electron density is cylindrically symmetrical in these triple bonded molecules.

H—C≡C—H
an alkyne

CH_3—C≡N
a nitrile

Delocalized π bonds

More than two adjacent p orbitals can combine to form a set of molecular orbitals where the electron pairs are shared by more than two atoms. These form 'delocalized' π bonds. The methanoate ion $HCOO^-$, for example, can be drawn as either

H—C(=O)—O⁻ or H—C(—O⁻)=O

but is better represented by H—C(O)(O) ⁻ where

shows the probable distribution of the two delocalized pairs of electrons over the three atoms. This explains the observed equal C–O bond lengths.

In the case of benzene, C_6H_6, there are six p orbitals making *three delocalized* bonding orbitals, each holding a pair of electrons. These are usually summarized as a double ring doughnut of electron density.

A single conventional structure does not adequately describe the methanoate ion. Instead of the 'cloudy' picture, the methanoate ion can be described using the two conventional structures. Neither of these contributing structures actually exists but the true structure of the methanoate ion lies in between them. This method of description is called 'resonance' and is denoted by double-headed arrows between the contributing structures, e.g.

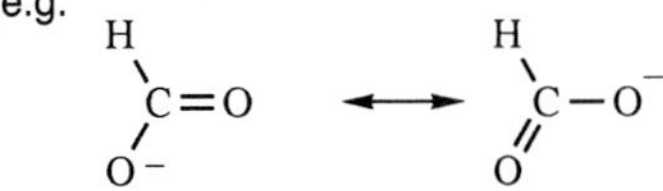

Resonance structures differ only in *electron* distribution; all the *atoms* stay in the same places.

Normal C—O and C=O bond lengths are about 0.14 and 0.12 nm respectively. Both carbon-to-oxygen bonds in the methanoate ion have a length in between these values: 0.13 nm.

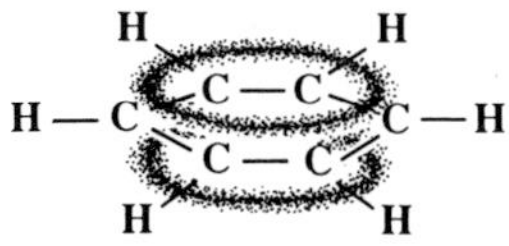

The σ bond framework of the molecule is shown as lines. The region of high electron density is above and below the plane of the ring. In structural formulae (see p. 7) benzene is represented as

either or or

The structure of benzene, and the evidence for it, will be discussed in more detail in Chapter 5 (see p. 67). The delocalization in benzene can be extended to other adjacent atoms. If delocalization is extended sufficiently

Compounds with a system of alternate single and double bonds are said to have *conjugated* double bonds.

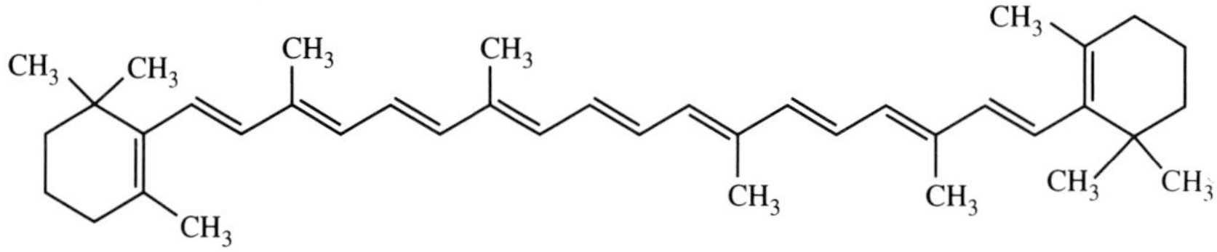

the compounds become able to absorb light energy in the visible region; they become coloured.

β-carotene: yellow colour in carrots and daffodils; E 160.

The dye INDIGO

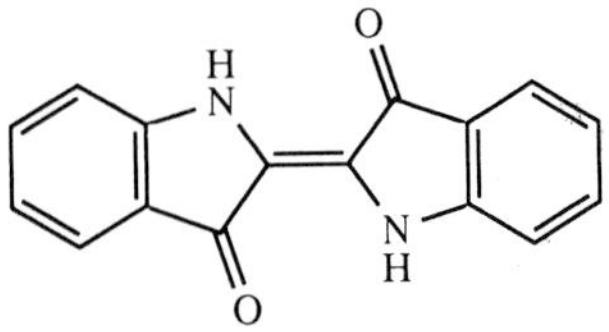

is obtained from a plant (an Indigofera species). A colourless sugar derivative is first isolated, which after hydrolysis is oxidized by air to the familiar blue dye used for jeans. The dye is not stable and gradually fades.

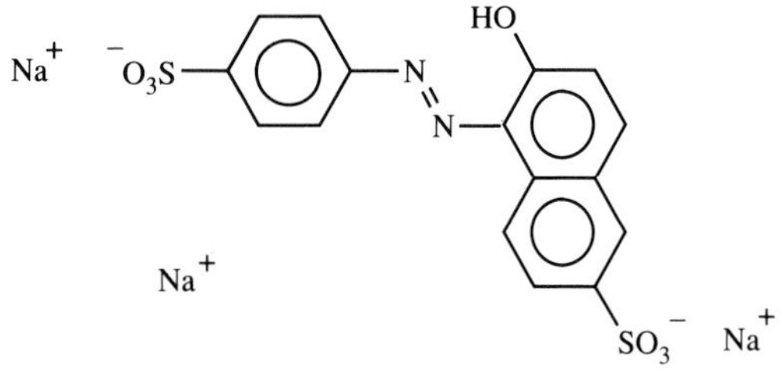

Sunset Yellow FCF; E 110.

Bond polarization: dipoles

Pairs of electrons in covalent bonds are only shared equally by atoms of the same element. The electron distribution is symmetrical.

C ⬬ C H ⬬ H

Dissimilar nuclei will have a differing attraction for the shared pair which results in an unsymmetrical distribution of electron density and a permanent slight separation of charge, a *dipole*. This dipole can be measured, and for a liquid its magnitude can be related to the solvent properties. H—Cl is actually H⬬Cl because the chlorine has a greater attraction for the shared pair than does the hydrogen. This gives a small separation of charge, the partial charges being shown as $\delta+$ and $\delta-$. The bond is said to be polarized. The presence of nonbonded pairs on only one of the atoms may also contribute to this dipole.

$\overset{\delta+}{\text{H}}$—$\overset{\delta-}{\text{Cl}}$

δ is used to mean 'a small amount of'; so that $\delta+$ means a fractional positive charge and $\delta-$ a fractional negative charge.

Chlorine is said to be more *electronegative* than hydrogen. Some other electronegativities are shown in Table 1.1.

Table 1.1 Electronegativities (Pauling scale)

H 2.1	C 2.5	N 3.0	O 3.5	F 4.0
	I 2.5	Br 2.8	Cl 3.0	F 4.0

The more electronegative of the two atoms in a dipole carries the $\delta-$. The bigger the difference in electronegativity of the two atoms, the bigger the dipole in a covalent bond between them. The molecule as a whole will have a measurable dipole moment, unless the dipoles within it cancel each other out by being arranged symmetrically.

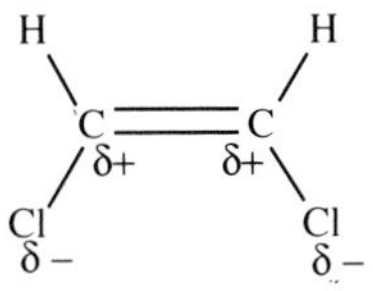

Permanent dipole

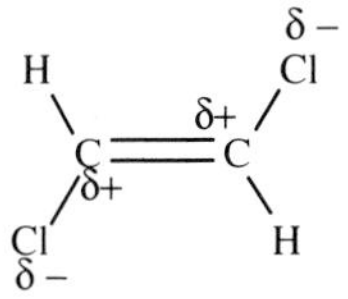

No permanent dipole

Bond polarization: inductive effects

The effect of a permanent dipole may be felt some distance away in a molecule, the dipole in one bond inducing a similar smaller dipole next door. Thus C—C—Cl is actually C◉C◉Cl and there is an induced dipole in the C—C bond. This can be written C⇸C⇻Cl; NB it is *not* a dative bond X⟶Y. A methyl group attached to carbon also has an inductive effect, H_3C⇸C (see right).

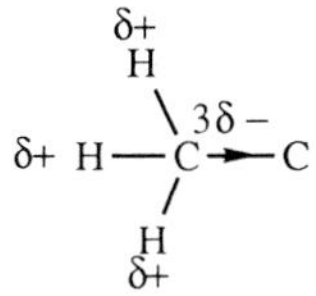

Inductive effects are most useful in supplying a rule of thumb 'explanation' for differences in reactivity.

Structural formulae

Molecules are drawn using structural formulae, a kind of shorthand for the full bonding diagram. Some examples will show how they are used; look at the yellow dyes on p. 6 and these simpler molecules below.

Methane: CH_4 or

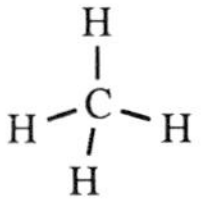

Propanone: CH_3COCH_3 or

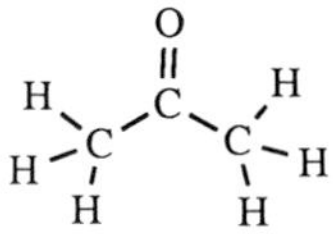

Ethanoic acid: CH_3COOH or

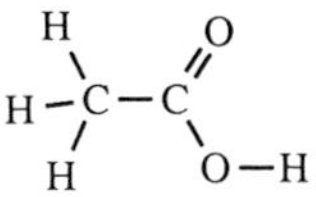

It is common to see the structures of alkyl groups drawn with apparent bond angles of 90°, but remember that the carbons are *tetrahedral* with angles of about 109°.

Each — represents one bond or a shared pair of electrons. The advantage is that we can see at a glance exactly which atoms are joined and by how many bonds, but we still have to remember details of *type* of bond (p. 2) and *shape* of molecule, which are not always shown explicitly.

For larger molecules (like the yellow dyes) showing every bond is cumbersome, and we usually restrict them to *functional groups* (see p. 8), grouping the other atoms together in such a way as to make the structure clear.

Propanonitrile would become $CH_3CH_2C \equiv N$

or even C_2H_5CN

Stearic acid, from animal fat, is then $CH_3(CH_2)_{16}COOH$.

For complex compounds even this convention would take a good deal of writing time, whilst drawing attention away from the functional groups which are the bits that really influence the chemistry.

Hexane can be shown as and cyclohexane as

Each corner or end represents a carbon atom and each line a bond. Hydrogen atoms on carbon are not shown and we assume that each carbon atom has the right number of hydrogens to give it four bonds altogether.

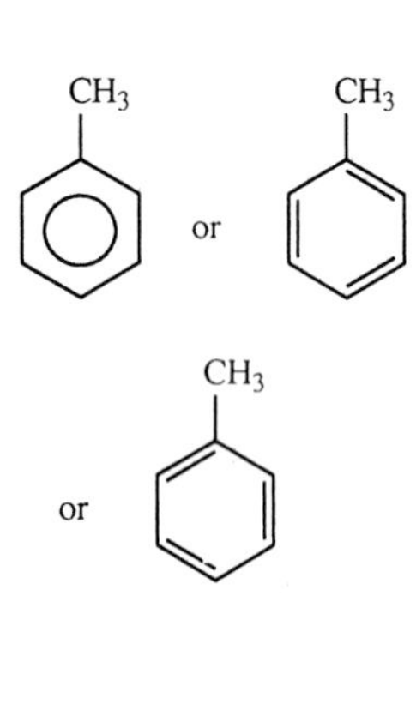

We have already shown, on p. 5, the representation of the delocalized system in benzene. So the structures on the left represent methylbenzene, where one of the hydrogens on the ring has been replaced by a CH_3 or methyl group. The versatility of these shorthand forms can easily be appreciated in the structures given below and the two yellow compounds on p. 6.

Cholesterol

Adenosine triphosphate (ATP)

Menthol

Name the functional group in MENTHOL, which is extracted from plants of the mint family and is used in sweets and toothpaste.

Functional groups

Compounds such as ethanol, CH_3CH_2OH, and propanol, $CH_3CH_2CH_2OH$, have very similar chemical properties because the 'business ends' of their molecules, where reactions occur most readily, are identical. They both have the group of atoms $-CH_2OH$, a primary alcohol.

In the same way, all carboxylic acids possess —COOH as the business end of their molecules. We have already come across several more of these *functional groups*.

OXETANOCIN is a natural product from a bacterium which is used clinically against HIV. It includes two primary alcohol groups (—CH_2OH), a four-membered ether ring, and a primary amine.

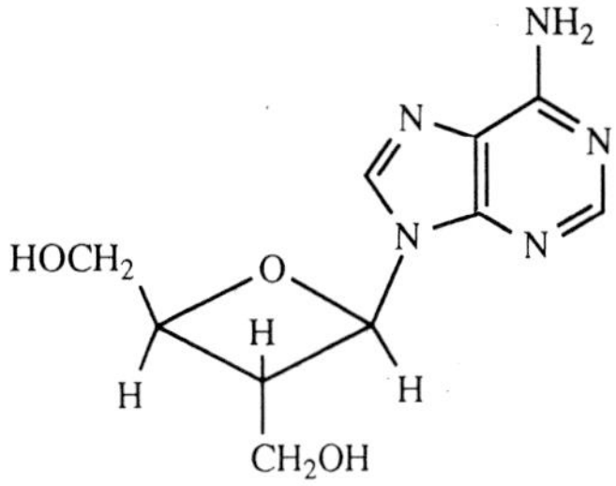

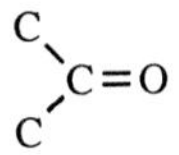

ketone

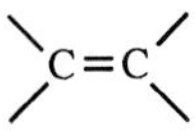

alkene

—C≡N

nitrile

—NH_2

primary amine

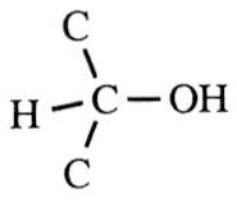

secondary alcohol

Part of any organic course involves learning the structures of these functional groups and some of the reactions associated with them. You will find this material in your notes or in any standard textbook. Such an approach tends to emphasize the differences in behaviour of the various functional groups. In this book we are more concerned with drawing together the common underlying themes involved in the structures of the compounds and the mechanisms of their reactions.

Dot diagrams

Dot diagrams give a simple way of following the formation of ions or molecules from atoms. You may be familar with this idea already. Outer shell electrons are represented by dots (or asterisks). The formation of sodium and chloride ions from their respective atoms would be shown as in the diagram on the right.

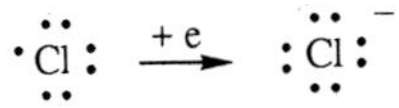

The electron pairs on the chloride ion are indistinguishable from each other. Covalent bonding in methane would be represented as

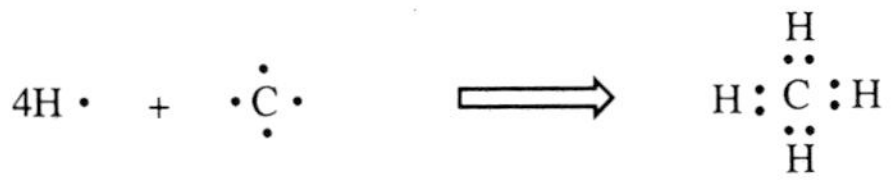

$$Na\cdot \longrightarrow Na^{+} + e$$

The use of such diagrams allows us to get the electron bookkeeping correct, with the formation of the right number of bonds. The use of dots allows us to keep track of electron pairs during the bond-making and -breaking in a reaction (e.g. p.21). It also helps us to understand why some of the intermediates of organic reactions are charged (see p.59).

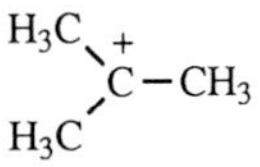

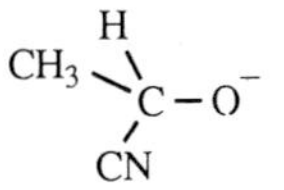

You should try drawing dot diagrams for these two ions.

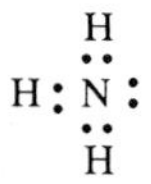
ammonia

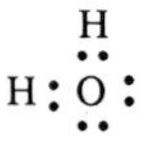
water

Nonbonded pairs

If we draw a dot diagram for ammonia, we find that the outer shell of the nitrogen atom contains a nonbonded (or lone) pair of electrons. The oxygen atom in the water molecule similarly has two nonbonded pairs. The molecular orbital picture for ammonia would have the nonbonded pair on top.

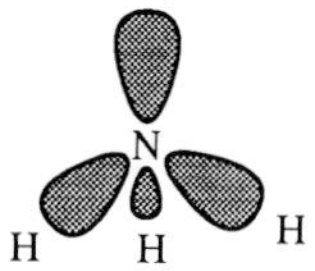

In a full structural formula, we would write

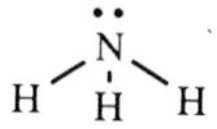

We can simply abbreviate this to :NH_3. In practice, the nonbonded pairs are often omitted in formulae such as NH_3, which can make it difficult to see which ions or molecules will act as bases (p. 33) or nucleophiles (p. 40) until one is familiar with them.

Nonbonded pairs can be used to form *dative covalent bonds* with other atoms or molecules. The 'type reaction' is H_3N: + BF_3 going to $H_3N \rightarrow BF_3$ where $\rightarrow$ represents the dative bond between N and B using the nonbonded pair on the nitrogen. This is an underlying theme in many organic reactions involving polar mechanisms (e.g. p. 41). Although we do not draw all nonbonded pairs onto structural formulae even when they are needed, you will have to remember where they are. Most nucleophiles have them.

COCCINELLIN is a 'defence compound' exuded by ladybirds when attacked.

The central N atom in coccinellin has used its nonbonded pair to make a dative bond to oxygen.

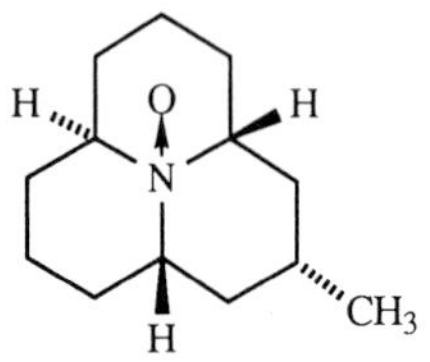

1.4 Bond strengths and lengths

Two factors influencing the strength of a bond are its type and the sizes of the atoms involved. We define bond strength as the enthalpy change of the process $X{-}Y_{(g)} \rightarrow X_{(g)} + Y_{(g)}$. The term *bond energy* is often used for this.

Table 1.2

Bond	Length (nm)	Bond energy ($kJ mol^{-1}$)	Type
C—C	0.154	346	σ
C=C	0.135	610	$\sigma + \pi$
C—O	0.143	358	σ
C=O (aldehyde)	0.122	736	$\sigma + \pi$

Table 1.2 shows that σ bonds are stronger than π bonds for C to C connections. The π bond between C atoms is therefore the easier part of the double bond to break and the chemistry of alkenes is dominated by this.

For carbon to oxygen double bonds the reverse appears to be true. This could be one reason why there are more addition/elimination reactions in C=O chemistry than in C=C chemistry (see p. 49).

Short bonds are usually strong ones. In the case of multiple bonds, the attraction of *two* shared pairs for the two positive nuclei pulls them closer together. The effect of atomic size on bond energy can be seen in the series C—halogen. The smaller the halogen atom, the closer the bonding nuclei can get to each other (Table 1.3).

Table 1.3

Bond	Length (nm)	Bond energy (kJ / mol^{-1})
C—F	0.138	452
C—Cl	0.177	339
C—Br	0.194	280
C—I	0.214	230

All other things being equal, one would expect the C—I bond to be the easiest of the four to break. It is the comparative difficulty of breaking C—F and C—Cl bonds that has made chlorofluorocarbons (CFCs) a menace to the environment. Compounds such as CCl_2F_2 linger for years in the atmosphere and are unaffected by bacteria. There is evidence that they damage the ozone layer. They are now banned in the UK.

Iodobutane is indeed hydrolysed faster than chlorobutane, partly because of the weakness of its C—I bond (see also p. 41).

It is their very lack of reactivity that makes them so useful as refrigerants, anaesthetics, and non-stick coatings.

1.5 Stereochemistry

Bond angles, molecular shapes

'All electron pairs, being similarly charged, are mutually repulsive but nonbonded pairs are more repulsive than others.' All we need in order to apply this principle to work out shapes is a knowledge of the number and type of electron pairs around the atom of interest. Their distribution in space will be such that they are all as far away from each other as possible. Their mutual repulsion leads to the shape involving the maximum separation of electron pairs. Dot diagrams (p. 9) make good starting points.

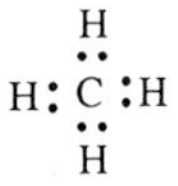

Methane has four bonding pairs. Maximum separation is achieved by the four pairs taking up a *tetrahedral* arrangement with bond angles of 109.5°. In the figure (right), thick lines ◀ represent bonds that are coming out of the page towards us, ııııı bonds that are going away from us into the page. Lines of normal thickness — are bonds in the plane of the page.

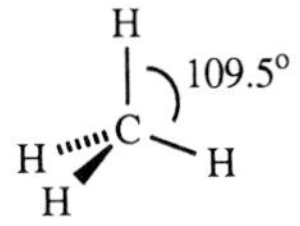

H:C:H (with H above and below)

109.5°

The same distribution applies to ammonia (three bonding pairs, one nonbonded pair), though the greater repulsive effect of the nonbonded pair, which stays closer to its parent nucleus than a bonded pair, reduces the bond angle to 107°. The actual *atoms* in ammonia are arranged in a pyramid. The angular structure of water can be rationalized in the same way with two nonbonded pairs.

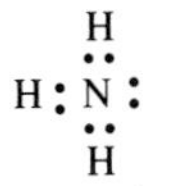

ammonia  107°

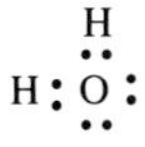

water 105°

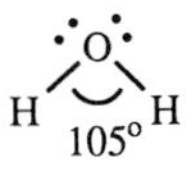

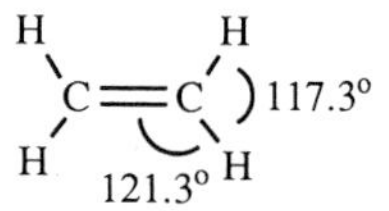

H H C=C 117.3° H 121.3° H

The two pairs of electrons in a double bond appear to be a little more repulsive than a single pair, and they can be counted as one region of electron density for shape purposes.

It is usually good enough to take these *trigonal planar* angles to be 120°. The need for overlap of the two parallel p orbitals to make the π bond means that the whole molecule is flat. Rotation around the central C=C bond is difficult energetically because it involves breaking the π bond; the overlap of the two p orbitals would be minimal if the two CH_2 ends were at right angles (see p. 4). The same trigonal planar carbon atoms are seen in benzene, in methanal, and in carbocations (e.g. p. 45). BF_3 is also a trigonal planar molecule.

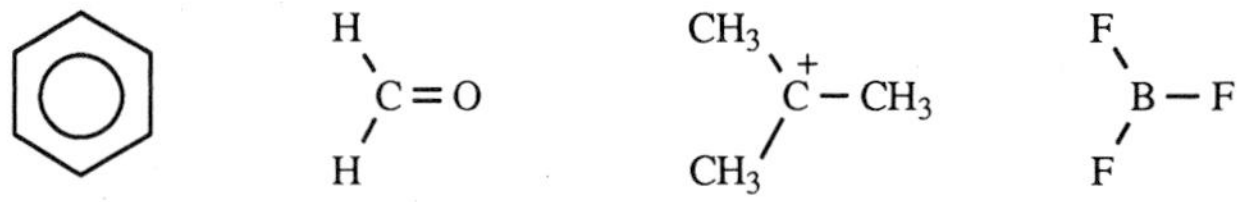

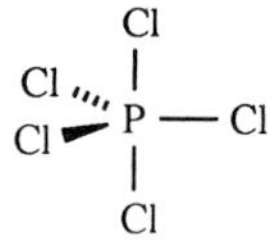

The other shape we are likely to come across is the *trigonal bipyramid* arrangement for which the classic example is gaseous PCl_5, where five pairs of electrons are associated with the central atom. For overall maximum separation three pairs of electrons are in a *trigonal planar* arrangement with the other two pairs at right angles to this plane. We will meet this shape again in the transition states for some nucleophilic substitution reactions (see p. 41).

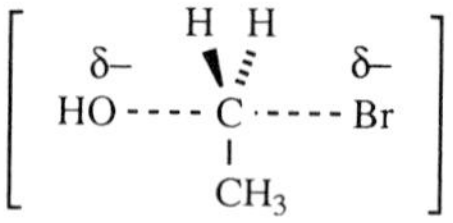

The C—O and C—Br bonds are drawn dotted to show that they are neither completely made nor completely broken but in between, as one might expect in a *transition state*. The carbon atom is *not* 5-valent!

Overall, this transition state must have one whole negative charge (see p. 28).

Structural isomerism

There are several ways in which the same set of atoms can be joined together to form different molecules. The simplest way is to join the atoms in a different order. Compare

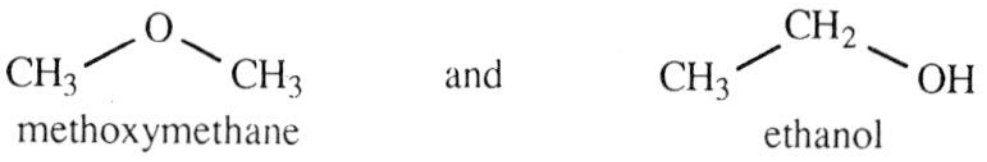

methoxymethane and ethanol

and the three possible nitromethylbenzenes.

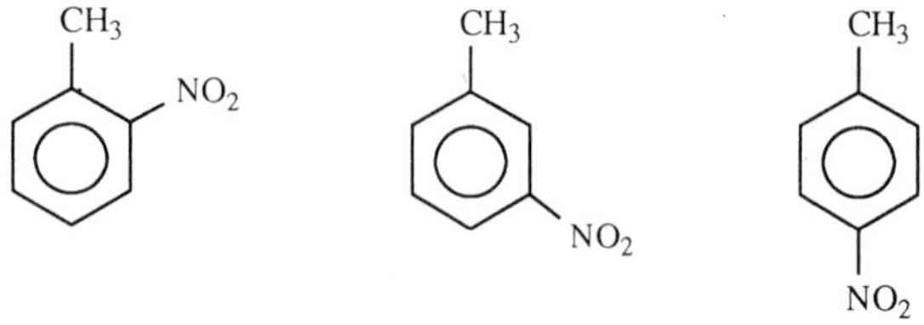

Pick out the carboxylic acid and ester groups in ASPARTAME, a synthetic 'sweetener' 150 times as sweet as sugar. Draw out the structure showing *all* the multiple bonds.

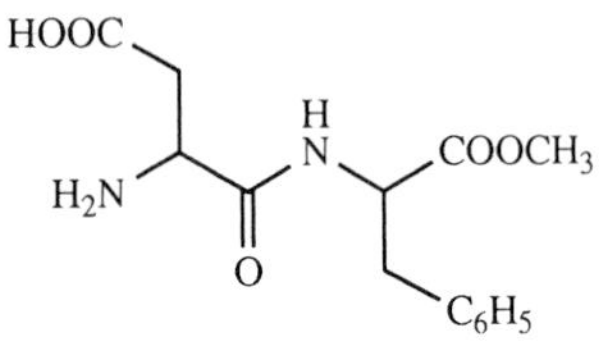

These are usually referred to as *structural isomers* or simply *isomers*. As well as having different physical properties such as melting and boiling points, these isomers are likely to have different chemical properties, very different if the functional groups have been changed. For example, both

CH_3CH_2COOH and CH_3COOCH_3 are $C_3H_6O_2$, but one is an *acid* and the other an *ester*.

Stereoisomerism: geometric (*cis/trans*) isomerism

Stereoisomerism describes isomerism where two molecules have the same atoms joined to each other in the same order but with a *different arrangement in space*. Such isomers exist because they lack symmetry.

The first kind is called *geometric* or *cis/trans isomerism*, with but-2-ene as an example. Written linearly, $CH_3CH{=}CHCH_3$, the molecule gives no hint of the two possibilities. But when the structure is drawn to show the shape, one can either write it with both the hydrogens attached to carbon on the same side of the double bond, the *cis* isomer, or on opposite sides, the *trans* isomer.

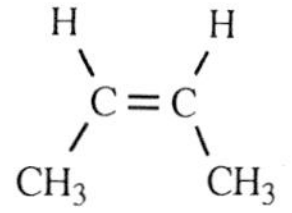

```
  H       H
   \     /
    C = C
   /     \
CH3       CH3
```

cis but-2-ene

```
  H       CH3
   \     /
    C = C
   /     \
CH3       H
```

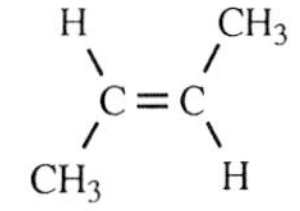

trans but-2-ene

There is restricted rotation about the C=C bond and the two isomers are not interconvertible without breaking the π bond and rotating one end. This is energetically unlikely. The two compounds have different physical properties, such as boiling point, and slightly different chemical properties. One might think that

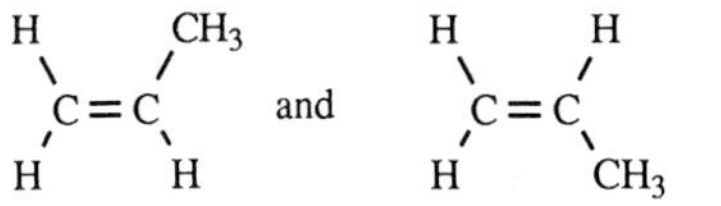

```
H     CH3           H     H
 \   /               \   /
  C=C       and       C=C
 /   \               /   \
H     H             H     CH3
```

are also isomers, but because the molecule as a whole is flat one can simply turn the left hand molecule over to get the right hand one. The key feature for this isomerism is the presence of *two different* groups on *each* end of the double bond. The conversion of one *cis* isomer into its *trans* relation is a central feature of human vision.

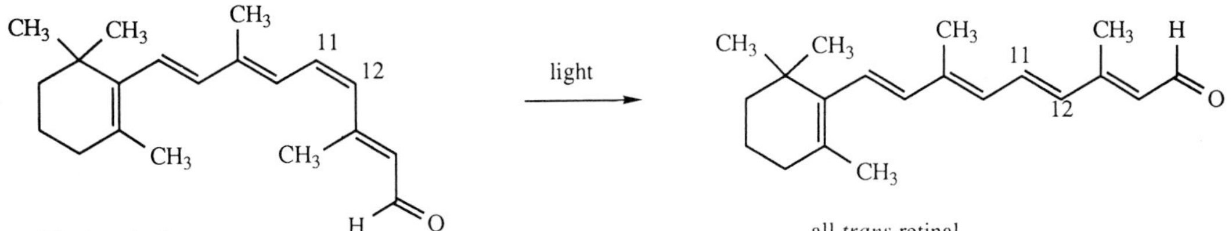

11-*cis*-retinal —light→ all *trans*-retinal

We can draw butane $CH_3CH_2CH_2CH_3$ showing three dimensions to see if a similar isomerism is possible. Remember that thick ◢ bonds are coming out of the page, ııııı bonds are going away from us, and normal lines — are bonds in the plane of the paper.

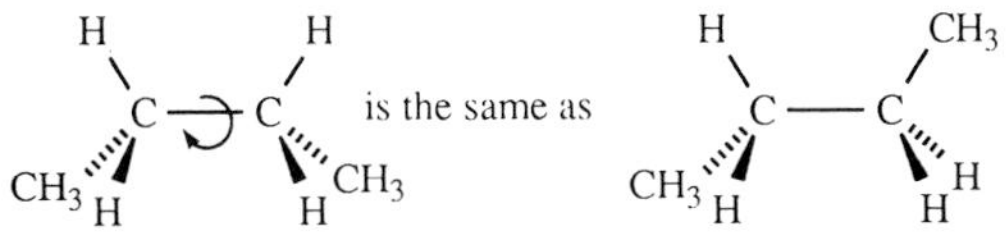

is the same as

In fact the easy rotation about the central single carbon to carbon σ bond means that the two drawings are of the same molecule, one end being rotated relative to the other by 120°. It is much easier to see these three dimensional shapes with molecular models, and it is well worth having regular access to a set of them. It could be said that Watson and Crick won their Nobel prize for playing with models of DNA.

ALANINE

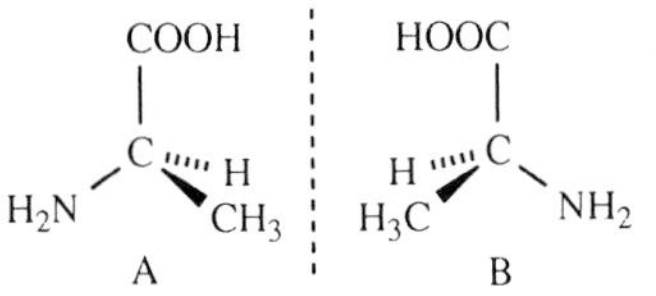

A　　　B

In diagram B you can see that we have drawn the groups the 'wrong' way round:

H_3C- for methyl, $-CH_3$;
HOOC– for carboxylic acid, –COOH.

The groups are not different at all, but just drawn backwards to emphasize the atoms which are bonded together and the mirror-image relationship of A to B. The amino group in diagram A is also reversed for the same reasons.

Stereoisomerism: optical isomerism

The second type of stereoisomerism is caused by a lack of symmetry in a molecule's structure so that the molecule and its mirror image are actually different, like a left hand and a right hand. No amount of twiddling, with models or on paper can make the mirror structures identical, without breaking bonds; they are *nonsuperimposable*. An example is the amino acid alanine (2-aminopropanoic acid).

If we try to rotate molecule A, hoping to get molecule B by overlapping the central C atom, the COOH and the NH_2 groups, we are unsuccessful because now the methyl (CH_3) group is going into the paper instead of coming out. Confirm this for yourself using molecular models. The isomers differ in the direction in which they rotate the plane of plane polarized light; otherwise their physical properties are the same. They are identical in chemical reactivity except in cases where shape and symmetry are critical, most notably in reactions with other optical isomers including enzymes (see p. 85). In simple cases we can recognize a molecule with optical isomers by the presence of one or more *asymmetric* carbon atoms, that is, a carbon atom bearing four different substituents. A few minutes spent playing with drawings or models will show you that the mirror image is different. On the other hand a little twiddling of

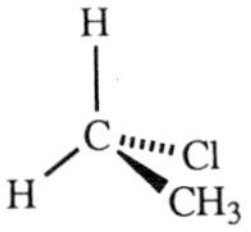

should show that the presence of two identical substituents on the carbon atom rules out optical isomerism. A molecule possessing asymmetry leading to optical isomerism is said to be *chiral*. Biological systems are often very selective about optical isomers. We can only metabolize amino acids shaped like B above. One optical isomer of limonene occurs in oranges, the other in peppermint oil, and both in turpentine. The two isomers smell different!

Only *one* optical isomer of DOPA, called L-DOPA, is effective in the treatment of Parkinson's disease.

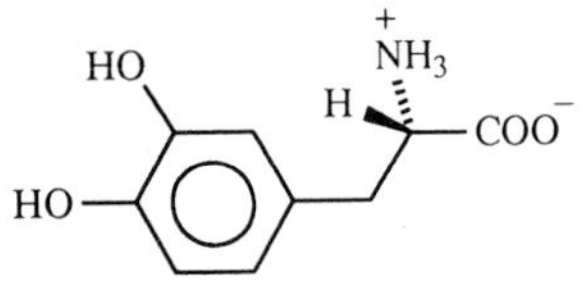

Only *one* isomer of MORPHINE acts as an addictive but efficient painkiller.

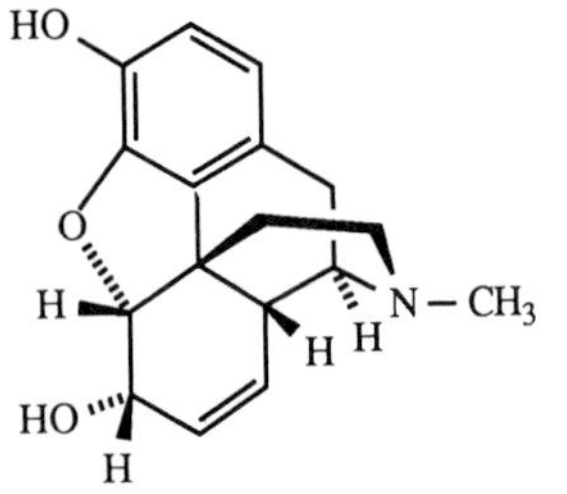

How many asymmetric carbon atoms are there in these molecules?

Limonene from lemons or peppermint oil

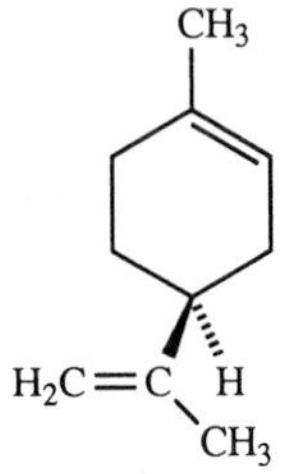

Limonene from oranges

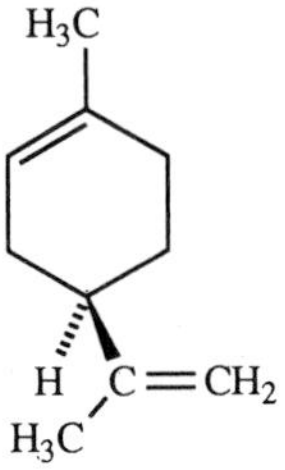

1.6 Intermolecular attractions

There are a number of ways in which separate molecules can attract each other. These affect physical properties such as melting point and boiling point as well as solubility.

Dipole–dipole attractions

The permanent dipoles (see p. 6) in molecules will attract each other, following the general principle that opposite charges attract, e.g. propanone.

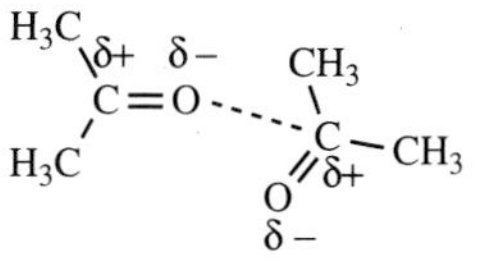

$(H_3C)_2C^{\delta+}{=}O^{\delta-} \cdots C^{\delta+}(CH_3)_2{=}O^{\delta-}$

This is a weak attraction (only a few kJ mol^{-1}; compare with the energy of covalent bonds in Table 1.2). This weak attraction is enough to account for the relative boiling points of propanone and butane, which have a similar number of electrons altogether. Remember that in going from liquid to gas it is only those bonds *between* molecules that are broken.

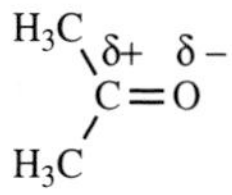

The attraction of molecules with permanent dipoles (*polar molecules*) for fully charged ions helps to explain the solubility of compounds like sodium chloride in water. The figure shows only one water molecule per ion for simplicity. There will actually be several clustered around each ion.

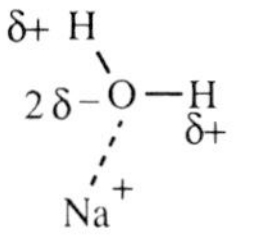

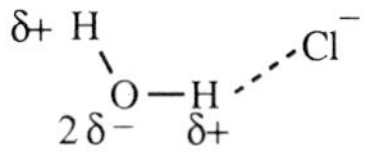

$H^{\delta+}-O^{2\delta-}(-H^{\delta+}) \cdots Na^+$ $\quad$ $H^{\delta+}-O^{2\delta-}-H^{\delta+} \cdots Cl^-$

This is *ion–dipole attraction*.

$(H_3C)_2C^{\delta+}{=}O^{\delta-}$

Propanone, b.p. 56°C permanent dipole

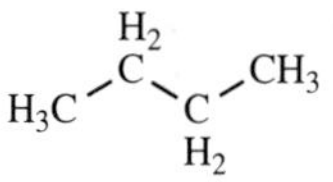

$H_3C-CH_2-CH_2-CH_3$

Butane, b.p. 0°C no permanent dipole

Van der Waals forces

A second type of weak attraction occurs between molecules which do not possess permanent dipoles. A rationale for this kind of attraction is that at any instant the electrons around even a symmetrical molecule may be unsymmetrically distributed, producing a temporary tiny dipole. This could induce a similar dipole in a neighbouring molecule. All these interactions give net attraction and are called *van der Waals forces*. The more electrons, the better!

The more 'contact' the attracted molecules can have with each other the stronger the forces are likely to be. The tiny dipoles can get close enough to interact in molecules which can lie easily side by side. So straight chain alkanes have higher boiling points than branched ones containing the same number of atoms (see Table 1.4).

Table 1.4

Compound	b.p. (°C)
pentane	26
2-methylbutane	18
2,2-dimethylpropane	10

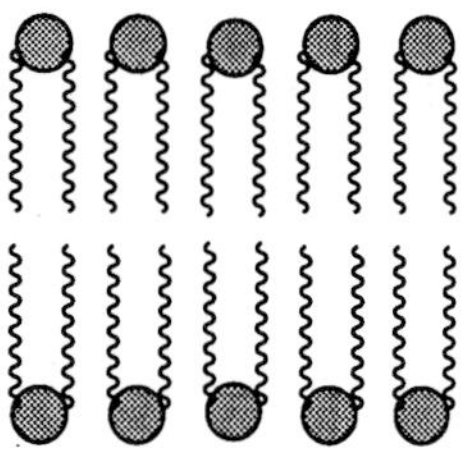

These van der Waals forces, again worth only a few kJ mol^{-1}, are generally weaker than dipole / dipole attractions. They are important in the maintenance of the coherence and fluidity of biological membranes. Phospholipid molecules with long hydrocarbon side chains are held together by van der Waals forces (see left).

They can also contribute to the tertiary structure of proteins. It is likely that at the active sites of some enzymes the catalysed reaction is essentially taking place in an organic medium, since all the water has been excluded.

Hydrogen bonding

A special case of dipole / dipole attraction involves hydrogen bonded to a small highly electronegative atom (fluorine, oxygen, or nitrogen). This H atom is attracted to another N, O, or F atom; this second atom must carry a nonbonded electron pair. This is known as *hydrogen bonding* and is worth 10–40 kJ mol^{-1}.

e.g. water

δ+ H
δ+ H \ O—H
\ ···· 2δ− δ+
2δ− O—H
δ+

The stronger the attractive interactions between molecules in a liquid, the more difficult it is for the molecules to escape from each other to form the vapour.

Hydrogen bonding is the strongest of the weak interactions and accounts for the high boiling points of alcohols compared with alkanes or ethers containing a similar number of electrons. Compare ethane (b.p. −89°C) $H_3C—CH_3$ with methanol (b.p. 64°C):

δ+ H
CH_3 \ O—CH_3
\ ···· δ−
O—H
δ− δ+

It also accounts for the solubility of alcohols and carboxylic acids (of relatively low molecular mass) in water (see p. 17).

Hydrogen bonding is essential for the complementary base pairing which holds together the double helix of DNA and forms the biochemical basis of genetics.

\δ− δ+ δ− δ+/
N—H ---- O=C
/ \

Hydrogen bonding is vital in the maintenance of secondary structure, the α-helices, and tertiary structure in proteins. It is involved in the binding of substrate molecules to the active sites of most enzymes, e.g. chymotrypsin (p. 17). The substrate is bound in by both hydrogen bonding and van der Waals attraction as shown.

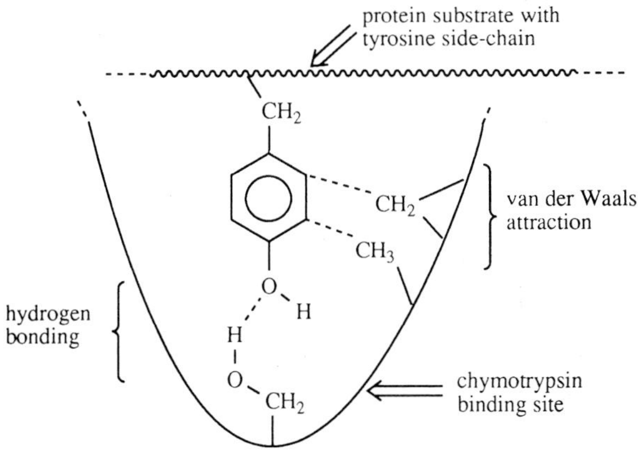

protein substrate with tyrosine side-chain

van der Waals attraction

hydrogen bonding

chymotrypsin binding site

PARACETAMOL is a widely used pain-killer. Draw diagrams to show how a molecule of paracetamol could be involved in hydrogen bonding with water molecules.

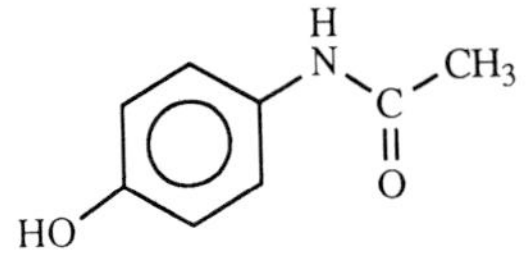

1.7 Solubility

This has been mentioned already in this chapter. Substances will mix completely, or a solute will dissolve in a solvent, as long as the sum of the attractions between molecules *after* mixing is equal to or more than that *before*. Where there is little change in the attractions during solution there will be no barrier to the mixing. This is the basis of the 'like dissolves like' principle.

Short chain alcohols and carboxylic acids, which are hydrogen bonded before mixing, will dissolve in water (also hydrogen bonded), with plenty of hydrogen bonding after mixing.

Propanone, with dipole / dipole attractions before mixing, will dissolve in water because of the possibility of *new* hydrogen bonds after mixing (see right).

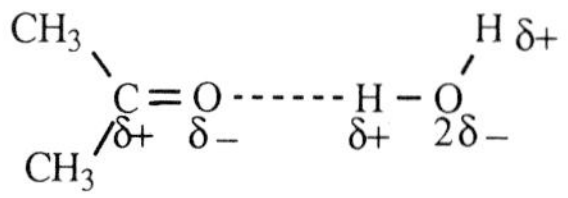

Compounds like hexane and benzene, whose molecules have no permanent dipoles, can mix freely because both possess van der Waals attractions before and after. But hexane, with van der Waals attractions, will not dissolve in water because, if it did so, some of the hydrogen bonding in the water would have to be sacrificed. Hexane molecules would physically get in the way with no compensating new attractive forces to make up for the loss.

Before: After:

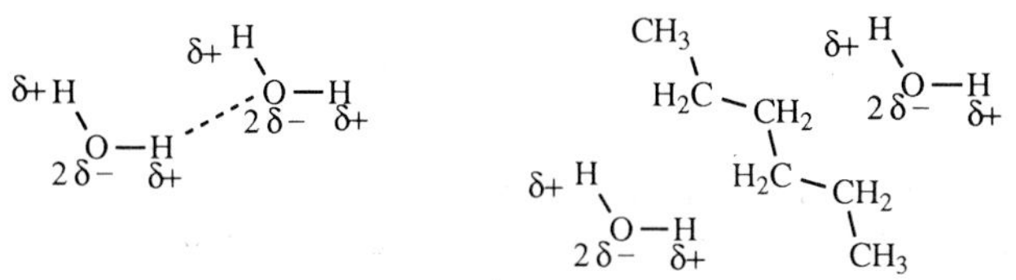

Even for alcohols and carboxylic acids in water, a balance is struck between the effects of the 'soluble' ends, $-CH_2OH$ (hydrogen bonding) and the 'insoluble' ends, $-CH_2-CH_2-CH_3$ (interference with hydrogen bonding). Alcohols and carboxylic acids with short hydrocarbon side-chains do dissolve in water, but as the chain length increases they become less and less soluble. The energy gain at the polar end is gradually outweighed by the loss at the lengthening nonpolar end.

The hydrogen bonds being broken are worth about -30 kJ mol^{-1}. Forming the extra new van der Waals attractive interactions, at about -4 kJ mol^{-1}, is not enough to make the whole process energetically favourable.

There is more to solubility than just these enthalpy changes, but the necessary consideration of entropy is beyond the scope of this book.

2 Mechanisms

2.1 Introduction

Organic reaction *mechanisms* are shorthand descriptions of how the starting compounds are made into the products. Molecules with the same functional groups usually react by similar mechanisms, and we can use mechanisms for known reactions to predict how other molecules will behave under similar conditions. Mechanisms are thus a connecting thread between similar reactions and allow us to *rationalize the products* of a known reaction and to *predict the reactivity* of other organic molecules.

Types of reaction

In this book we will be considering the mechanisms of three types of reaction: substitution, addition, and elimination.

Substitution reactions involve exchanging one atom or group of atoms on the organic molecule for another.

$$\text{e.g.}\quad CH_4 + Cl_2 \longrightarrow CH_3Cl + HCl$$

A chlorine has replaced one of the hydrogen atoms on the methane.

Addition reactions mean just that; one molecule adds onto another and nothing is lost.

$$\text{e.g.}\quad CH_2{=}CH_2 + Br_2 \longrightarrow CH_2BrCH_2Br$$

Elimination reactions involve the loss of a relatively small molecule from the organic reactant.

$$\text{e.g.}\quad CH_3CH_2OH \longrightarrow CH_2{=}CH_2 + H_2O$$

Water has been eliminated from the alcohol. Remember that elimination is the reverse of addition.

Heterolytic and homolytic bond breaking

Most reactions involve *bond breaking* as well as bond making. This can be of two types, heterolysis and homolysis.

In *heterolytic* bond breaking the pair of electrons in the bond becomes associated with only one of the atoms involved.

$$X{:}Y \longrightarrow X{:}^- + Y^+$$

The resulting fragments are likely to be charged. If so, they are ions.

In *homolytic* bond breaking each fragment retains one of the bonding pair of electrons. These fragments are called radicals.

$$X{:}Y \longrightarrow X\cdot + \cdot Y$$

2.2 Nucleophiles, electrophiles, radicals

It is useful to classify the chemicals with which organic molecules react. Leaving acid/base and some redox reactions on one side for the time being, we find that there are three types of reagent to deal with.

Most *nucleophiles* are molecules or ions with a nonbonded pair of electrons (see p. 10), with which they can form a new dative bond. They are 'electron rich'. Some examples are:

$H_2\ddot{O}$ $:\ddot{O}H^-$ $:\ddot{Br}:^-$ $:CN:^-$ $:NH_3$ $CH_3CH_2\ddot{N}H_2$ $CH_3CH_2\ddot{O}H$

We can look at the reaction of the triphenylmethyl cation with hydroxide ion, following the nonbonded pair.

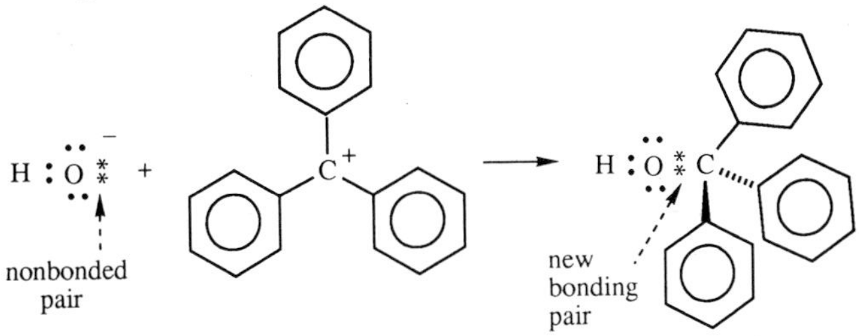

H : O ⁑⁻ + $(C_6H_5)_3C^+$ ⟶ H : O ⁑ $C(C_6H_5)_3$

nonbonded pair

new bonding pair

It is a common mistake to suppose that nucleophiles must be negatively charged, forgetting that the critical thing that many of them have in common is a *nonbonded pair*. Textbooks and papers do not help the beginner by leaving them out most of the time, assuming that everyone knows they are there. We shall do so ourselves in due course, but you need to remember where the nonbonded pairs are when you write mechanisms.

Electrophiles represent the other side of the coin. They are molecules or ions which are prepared to form a new covalent bond using a pair of electrons provided by another atom or molecule, often the organic reactant. Some examples are shown in the margin.

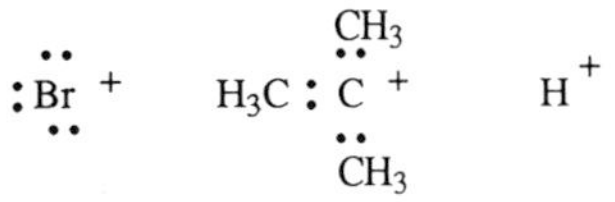

Sometimes the electrophiles, such as the three positively charged examples, actually have an *empty* bonding orbital available at low enough energy to accept the incoming pair of electrons. Uncharged electrophilic molecules are either polarized, e.g. HBr, or can be so readily, e.g. Br_2.

$:\ddot{Br}^+$ $H_3C:C^+(CH_3)_2$ H^+

$\overset{\delta+}{H}—\overset{\delta-}{Br}$ Br—Br

The addition of a proton, H^+, to an alkene will show how the first kind operates. The π electron pair on the alkene is shown as asterisks (*) for emphasis.

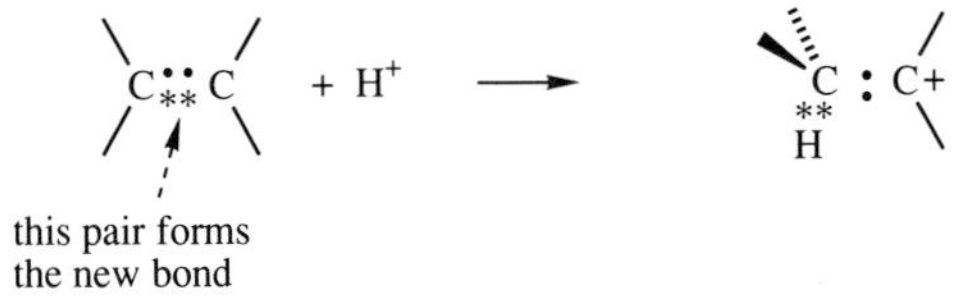

C::C (**) + H^+ ⟶ H**C : C+

this pair forms the new bond

In other cases heterolytic bond breakage in the electrophile makes an empty orbital available.

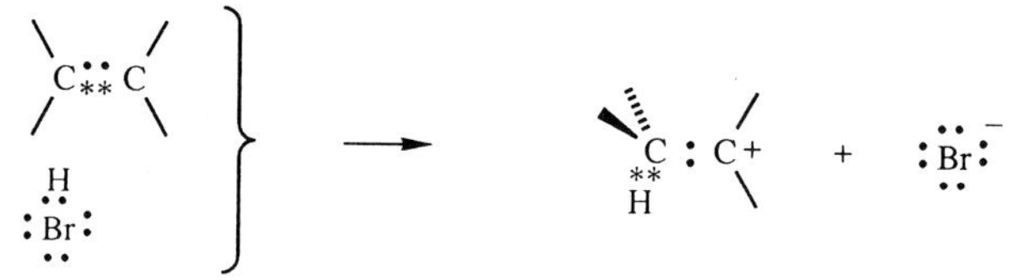

The π electron pair forms a new bond to the hydrogen atom while the HBr bond breaks heterolytically (see p. 18). The bromine atom of HBr becomes a bromide ion, carrying both of the original shared pair of electrons.

•Cl •CH_3 •Br

A *radical* carries an *unpaired electron*. Radicals are usually written with a dot to show these unpaired electrons.

It is worth drawing a full dot diagram for each of these to check the electron bookkeeping, and to show that they are electron deficient.

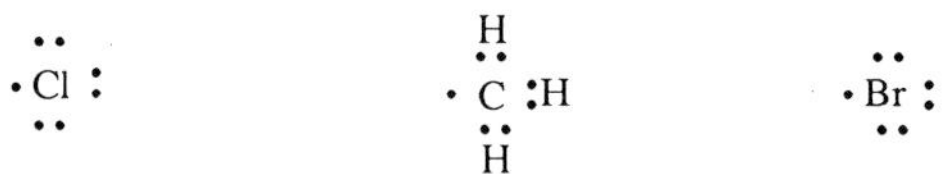

2.3 Drawing mechanisms using dot diagrams and curly arrows

Curly arrows and fishhooks are the conventional ways of depicting electron movements for pairs of electrons and single electrons respectively. We use for *ionic* (or polar) reaction mechanisms, that is, those involving nucleophiles and electrophiles. Bond breakage in these reactions results in the bonding pair of electrons remaining associated with only one of the atoms which were originally bonded so this is *heterolytic* bond breakage.

Br : C(CH_3)$_3$ → Br :$^-$ + C^+(CH_3)$_3$

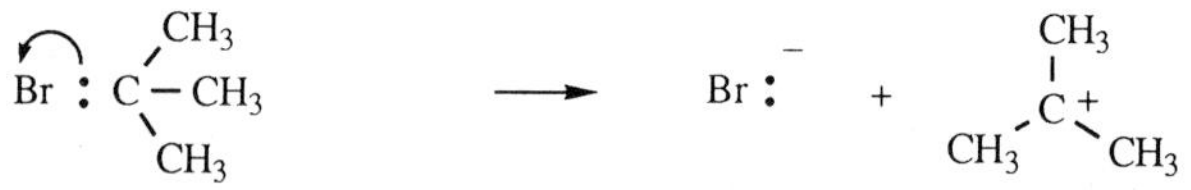

The pair of electrons in the breaking C—Br bond go to the bromine atom, forming a bromide ion. The positively charged carbon is short of electrons (electron deficient). The C—Br bond has broken heterolytically.

Curly arrows can be used for bond breaking, bond making, and for both making and breaking together.

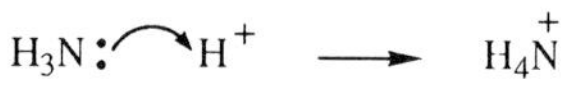

Examples: bond breaking $H-Cl \longrightarrow H^+ + Cl^-$

bond making $H_3N: \;\; H^+ \longrightarrow H_4\overset{+}{N}$

both $H_3N: \;\; CH_3-Cl \longrightarrow H_3\overset{+}{N}-CH_3 + Cl^-$

In each case a curly arrow must *start* from an electron pair, either a bonding or nonbonded pair, and *end* on an atom or in a new bond.

We use ⌒ for radical reactions, where it represents the movement of a single electron.

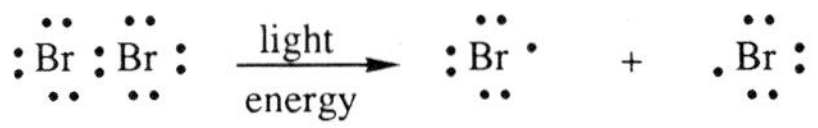

$$:\ddot{Br}:\ddot{Br}: \xrightarrow[\text{energy}]{\text{light}} :\ddot{Br}\cdot + \cdot\ddot{Br}:$$

or:

$$Br:Br \xrightarrow[\text{energy}]{\text{light}} Br^{\bullet} + {}^{\bullet}Br$$

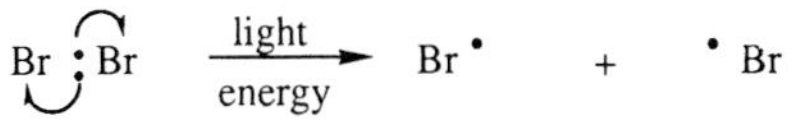

In radical reactions bond breakage leaves one of the bonding electrons on each of the originally bonded atoms. This is *homolytic* cleavage. Again, ⌒ can be used for making as well as breaking bonds.

We can show both types, ionic and radical, using full dot diagrams. It is probably a good idea to draw each mechanism in both ways until it becomes familiar.

Here is a *nucleophile* in action.

Dot version:

$$H:\ddot{O}:^{-} + H:\overset{CH_3}{\ddot{C}}:\ddot{Br}: \longrightarrow H:\ddot{O}:\overset{CH_3}{C}:H + :\ddot{Br}:^{-}$$

By emphasizing the electron pairs involved with different coloured pens we can follow what is going on. This is for our convenience; the four pairs in the bromide ion outer shell are quite indistinguishable from each other. Compare this with the shorthand curly arrow version.

Curly arrow version:

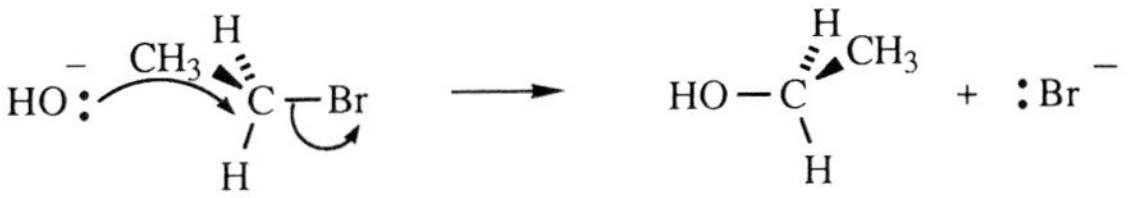

$$\overset{-}{HO}: + CH_3(H)(H)C{-}Br \longrightarrow HO{-}C(H)(H)CH_3 + :Br^{-}$$

You will frequently find such mechanisms drawn without the nonbonded pair on the hydroxyl ion. This nonbonded pair is *going to* form the new bond, a dative one, to carbon while the C—Br bond breaks heterolytically with the bonding pair of electrons *going off* on the bromide ion as it leaves. Curly arrows show *electron pairs going places*.

An *electrophilic* reaction is shown below. Notice the asterisk pairs.

Dot version:

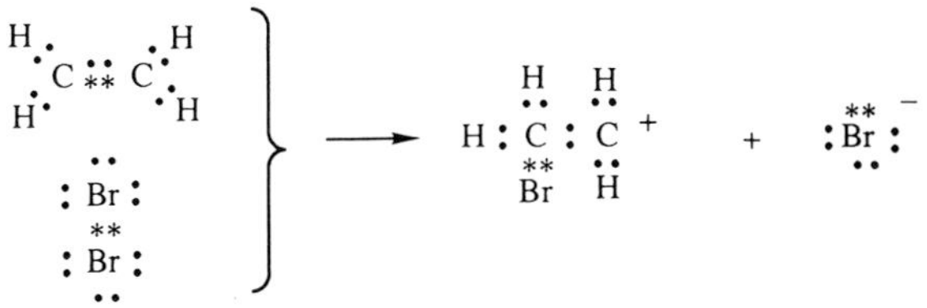

$$\left.\begin{matrix} H_2C{::}CH_2 \\ :\ddot{Br}: \\ ** \\ :\ddot{Br}: \end{matrix}\right\} \longrightarrow H:\overset{H}{\ddot{C}}:\overset{H}{\ddot{C}}^{+} + :\overset{**}{Br}:^{-}$$

In this reaction the right-hand carbon atom of ethene loses a half-share in the asterisk pair of electrons. Since formally it loses *one* electron from an uncharged state, the carbon must become positively charged.

The advantage of this method is that it helps us to see clearly why the right-hand carbon atom in the carbocation carries a positive charge.

Curly arrows are also used to relate resonance structures to each other, e.g. for the methanoate ion (see p. 5):

H—C(=O)—O⁻ ⟷ H—C(—O⁻)=O

and for benzene:

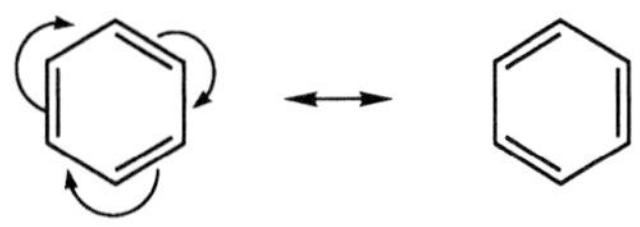

Curly arrow version:

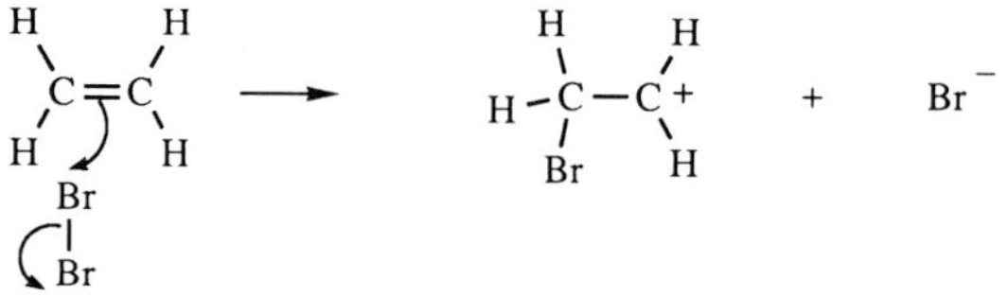

$H_2C{=}CH_2$ + Br—Br ⟶ H—C(H)(Br)—C⁺(H)(H) + Br^-

In the same way we can use either dot or fishhook versions of *radical* mechanisms, as for example in the reaction of methane with a chlorine radical.

Dot version:

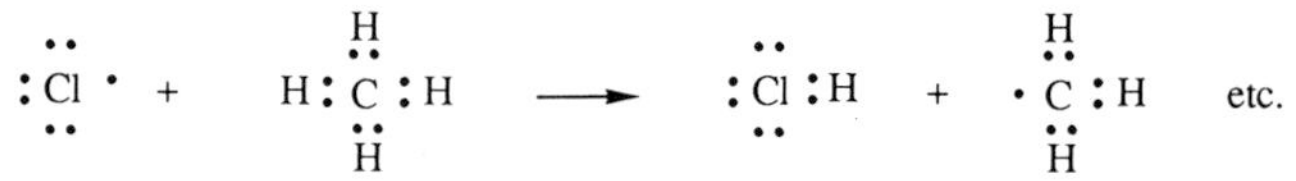

:Cl· + H:C:H (with H above and below) ⟶ :Cl:H + ·C:H (with H above and below) etc.

Fishhook version:

Cl· + H—C(H)(H)—H ⟶ Cl—H + ·C(H)(H)—H etc.

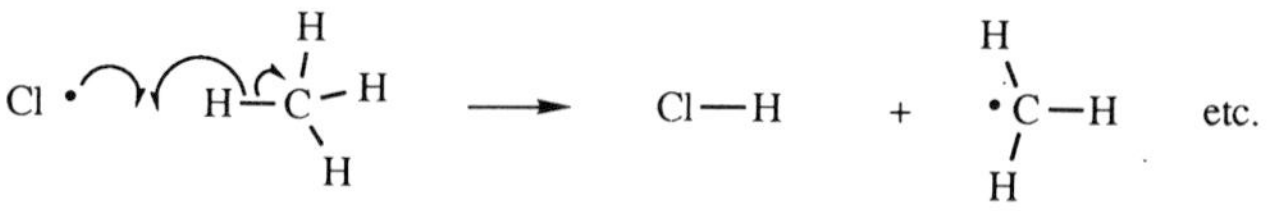

Curly arrow and fishhook mechanisms are simple descriptions. They need to be based on real experimental evidence, such as kinetics, isolation of intermediates and radioactive tracer experiments. Like all good theories they must be firmly based on fact.

2.4 Introduction: equilibria and rates

For all chemical reactions there are two vital questions to be answered.

1. How much product can be made?

2. How fast will it be made?

'How much' is determined by the *equilibrium constant* for the reaction at a particular temperature. This is related to what are called *thermodynamic* factors (see p. 23).

'How fast?' is determined by the rate constant at a particular temperature and what we call *kinetic* factors (see p. 24).

Organic chemists are interested in the answers to both questions because they wish to obtain the maximum yield of a desired product, as fast as possible. Some idea of how the bond breaking and bond making happens would be useful. This is the heart of the reaction mechanism, which must fit the observed facts about how much and how fast.

How much can be made? Equilibrium and energy profiles

Equilibrium. Some reactions seem to go to completion. This means that when the dust settles there are only products to be seen.

e.g. $$CH_3COCl + H_2O \longrightarrow CH_3COOH + HCl$$

In other reactions both starting materials and products are there at the end. The classic organic example is acid-catalysed esterification.

$$CH_3COOH + C_2H_5OH \overset{H^+}{\rightleftharpoons} CH_3COOC_2H_5 + H_2O$$

In fact the reaction is still going on at the apparent end, but with equal rate in each direction. There is no further change in concentration of products or reactants once the reaction has reached this *dynamic equilibrium*. The concentrations of products and reactants at this point are related by the *equilibrium constant*, K_c, which is constant at any particular temperature. For this example

$$K_c = \frac{[CH_3COOC_2H_5]\,[H_2O]}{[CH_3COOH]\,[C_2H_5OH]}$$

$$= 0.26 \text{ at } 373\,K$$

The [ ] terms refer to concentrations in mol dm^{-3}.

A high value of the equilibrium constant, K_c, would show that there will be a high proportion of products in the equilibrium mixture. A low value would indicate a low proportion of products at equilibrium.

Energy profiles. A simplified energy profile for a reaction shows the difference in energy between the reactants and products.

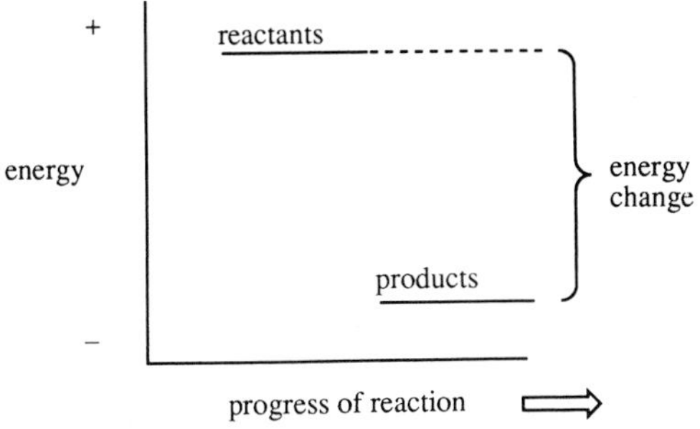

This reaction is thermodynamically favourable.

The 'energy change' for a reaction is related to its equilibrium constant by the expression

$$\text{Energy change} = -RT\ln K$$

where R is the gas constant ($8.3\ \mathrm{J\,K^{-1}\,mol^{-1}}$) and T is the temperature in kelvin. We can see that a large energy drop (−) in going from reactants to products will correspond to a high value of the equilibrium constant K (greater than 1) and the reaction *can* give a good yield of product.

This good yield can only be realized if the reaction is fast enough (see next section).

An energy increase (+), on the other hand

This reaction is thermodynamically unfavourable.

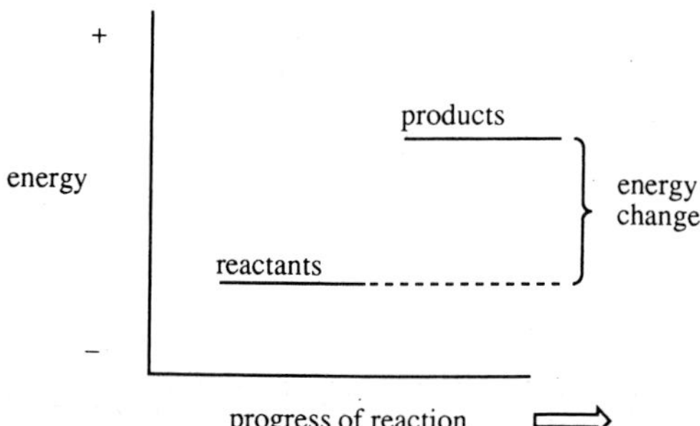

will give a low value of K (less than 1) and there will be a low proportion of product in the equilibrium mixture.

In a special case there is no energy change. What do you think would be the implications for equilibrium constant and yield of product in this case?

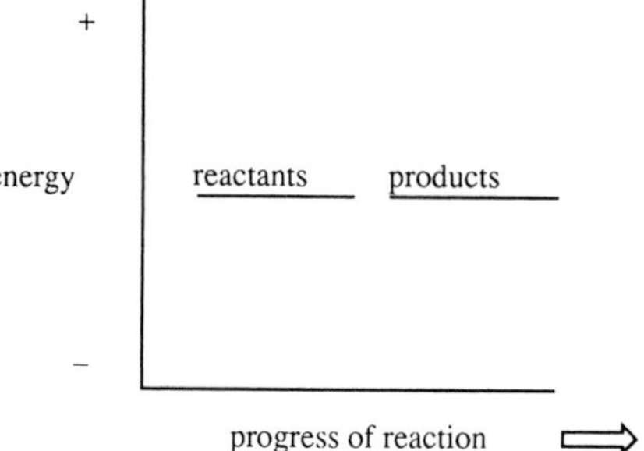

For many reactions in solution the *enthalpy changes*, ΔH, can act as a guide to the energy changes and, hence, to where the equilibrium lies. But they are only a *rough* guide.

How fast? Activation energy

Even reactions with large values of K are not instantaneous; they happen at varying rates. What is involved in two molecules reacting?

1. They must meet or collide.

2. They must presumably approach each other lined up so that the bond breaking and making process is made easy.

3. They must, between them, possess enough energy to get the reaction started.

A pen and its top can collide with each other from all directions.

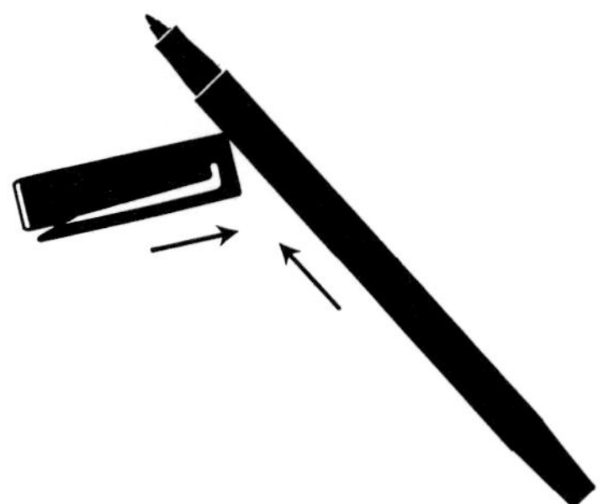

There is only one orientation which lets the top slip on, and they only stay together if it is pushed on with enough *energy*.

For chemical reactions, the energy barrier is known as the *activation energy*, E_a. The height of the barrier determines how fast a thermodynamically favourable reaction will go. Look at the reaction profiles below; the overall energy change is the same in each case.

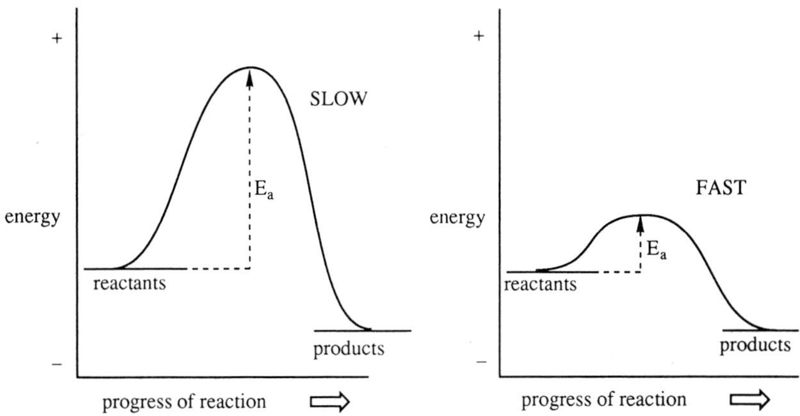

Note that the backward reaction, products ⟶ reactants, has a different (and higher) activation energy in both these cases.

A low value of E_a will normally correspond to a rapid reaction, a high value of E_a to a slower one. A reaction may be extremely favourable in overall energy terms, but very slow at low temperatures. For example, the equilibrium constants for the oxidation of glucose or magnesium are both large, but these substances are remarkably unreactive to oxygen at room temperature. The reaction rates are very slow because the activation energies are high.

There are two ways of speeding reactions up. *Firstly* we can heat the reactants so that a higher proportion of them have the activation energy on collision. This will give a higher proportion of successful collisions and therefore a faster reaction. This can be seen on the Maxwell–Boltzmann diagram (overleaf) which shows the distribution of kinetic energy in molecules at two different temperatures.

Maxwell–Boltzmann diagram to show the distribution of kinetic energy at temperatures T_1 and T_2.

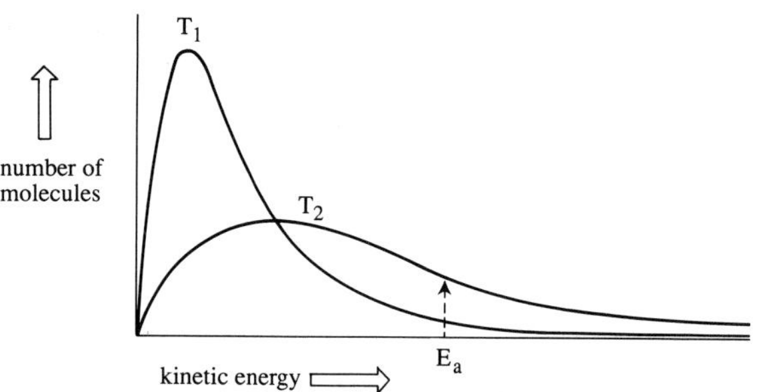

You can see that at T_2, the higher temperature, a higher proportion of molecules has the activation energy, E_a, or greater.

Secondly we can add a *catalyst* to the reaction mixture. The catalyst provides an alternative reaction pathway, involving a different set of bond breakings and makings, which has a lower activation energy. See p. 29.

Notice that the *overall energy change and the equilibrium constant, K, are unchanged*. Looking back to the Maxwell–Boltzmann diagram above, we can see that if we lower E_a we will increase the proportion of molecules with the activation energy or greater, without changing the temperature.

If we wish to oxidize glucose completely in the laboratory we can heat it in air until it burns.

$$C_6H_{12}O_6 + 6O_2 \longrightarrow 6CO_2 + 6H_2O$$

However, every cell in our bodies can achieve this oxidation, at *body temperature*, through a series of reactions, each one of which uses a specific catalyst or enzyme. Another example is:

$$2H_2O_2 \longrightarrow 2H_2O + O_2$$

Both manganese(IV) oxide and the enzyme catalase in liver will make the decomposition of hydrogen peroxide solution go faster, but catalase does it about a million times faster than manganese(IV) oxide.

Kinetics and rate equations. Rate determining steps

Looking at kinetics shows us which factors influence the rates of chemical reactions. We have already seen qualitatively how increasing temperature can raise the rate of reaction by increasing the proportion of molecules with enough energy to react—the activation energy (see p. 24). In the same way reactant molecules are more likely to collide if there are more of them about in a given volume. One would expect more *concentrated* reactants to react faster. The exact dependence of the rate of a reaction on the concentration of the reactants is found *by experiment* and is expressed in a *rate equation*.

$$A + B \longrightarrow C + D$$

k is the rate constant for the reaction.

Initial rate is proportional to $[A]^x[B]^y$: Rate $= k[A]^x[B]^y$

Here x and y are powers that are found *by experiment*. They do not necessarily bear any relationship to the balanced chemical equation at all, in marked contrast to equilibrium constants.

Let us take as an example the reaction

$$CH_3COCH_3 + I_2 \longrightarrow CH_2ICOCH_3 + HI$$

With acid catalysis the rate equation is found to be

$$\text{Rate} = k[CH_3COCH_{3(aq)}][H^+_{(aq)}]$$

Reaction rates are usually expressed as rates of change of concentrations.

The iodine does not appear in the equation at all. With all the powers in we would write

$$\text{Rate} = k[CH_3COCH_{3(aq)}]^1[H^+_{(aq)}]^1[I_{2(aq)}]^0$$

The reaction is then said to be *first order* with respect to propanone, *first order* with respect to hydrogen ions, and *zero order* with respect to iodine. It is *second order* $(1 + 1 + 0)$ overall, because the sum of all the powers in the rate equation is 2.

The proportionality constant k in the rate equation is known as the *rate constant* and is a characteristic of a particular reaction at a given temperature.

The mathematical relationship between E_a, the temperature T, and the rate constant k for a reaction is given by the Arrhenius equation:

$$k = Ae^{-E_a/RT} \quad (R = 8.3\ \text{J K}^{-1}\,\text{mol}^{-1})$$

Increasing T makes E_a/RT smaller and therefore $e^{-E_a/RT}$ gets larger, and k gets larger.

The exponential factor, $e^{-E_a/RT}$, is the fraction of molecules possessing at least the activation energy E_a. This corresponds to the area to the right of the E_a vertical line on the diagram on p. 26.

A is the frequency factor. It can be expressed as the product PZ, where P is a steric factor and Z is the total number of collisions per second of the reactant molecules.

You can now see how the Arrhenius equation relates to our pen-and-top analogy at the beginning of the section.

For many reactions it is possible to devise experiments which allow us to *deduce* the experimental rate equation. What does it tell us?

The actual sequence of bond breaking and making in the reaction is likely to take place in a number of stages, some fast and others slow. The slowest stage will act as a bottle-neck for the reaction and dictates the overall rate, so it is called the *rate-determining step*.

The substances involved *in or before the rate-determining step* will appear in the rate equation for the reaction.

For example, the nitration of 1,3,5-trimethylbenzene has first order kinetics.

1,3,5-trimethylbenzene (ring positions 1–6; CH_3 at 1, 3, 5) $+ HNO_3 \longrightarrow$ 2-nitro-1,3,5-trimethylbenzene (CH_3 at 1, 3, 5; NO_2 at 2) $+ H_2O$

The rate is proportional to $[HNO_3]$, but is *independent* of the concentration of trimethylbenzene. These data fit a mechanism in which the formation of the nitronium ion, NO_2^+, from nitric acid is the slow or rate-determining step. The rate of the slow step dictates the rate of reaction. Nitration is discussed further on p. 70.

If we understand something of the kinetics and mechanism of a reaction, we may be able to see how to increase the rate of a slow stage to give us more of the desired product.

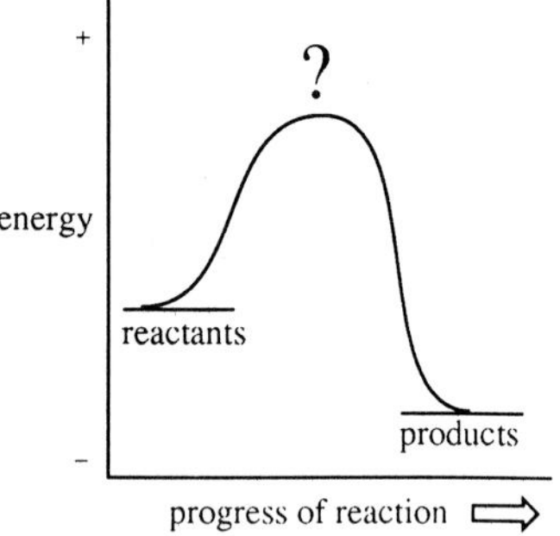

Intermediates and transition states

What exactly is the activation energy? What is happening at the top of the energy hump? In the substitution reaction

$$CH_3CH_2CH_2CH_2Br + OH^- \longrightarrow CH_3CH_2CH_2CH_2OH + Br^-$$

the C—Br bond begins to break at the same time as the C—OH bond begins to form.

There comes a point when the C—Br bond is weakened and the new C—OH bond is only partly formed. At this point we are poised on an energy maximum from which it is downhill in any direction, either back to C—Br and OH^- or on to C—OH and Br^-. The energy we have invested so far is the *activation energy, E_a*. We refer to this maximum energy (downhill either way) state as the *transition state*.

There is an overall charge of −1 on the transition state, which is shown with two partial negative charges (δ−) in the diagram. See p. 12 for the shape of TS.

$$HO^- + H_2C(C_3H_7)\text{—}Br \longrightarrow \left[\overset{\delta-}{HO}\text{----}C(H)(H)(C_3H_7)\text{----}\overset{\delta-}{Br} \right] \longrightarrow HO\text{—}C(H)(H)(C_3H_7) + Br^-$$

The TRANSITION STATE, TS

Transition states cannot be detected. We presume they are there because of the energy barrier.

You can create a model for this by partially inflating one of those 'sausage' balloons with separate panels as in (1).

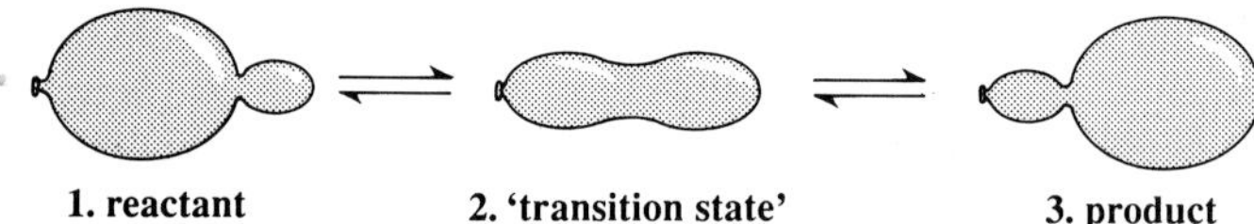

Now squeeze the air into the second panel. Part-way through this process you will suddenly feel that the balloon is poised (2). A tiny extra squeeze sends the air on into the next panel (3); a slight relaxation lets it back into the first (1). This in-between position (2) is like the transition state. If you let the balloon go, the air always goes to one end or the other. But the balloon has got to go through the 'transition state' if the air is to be shifted from one end to the other.

Now we can extend this idea to a balloon with three panels, the *reaction* being to get the air from the left hand panel into the one on the right. You

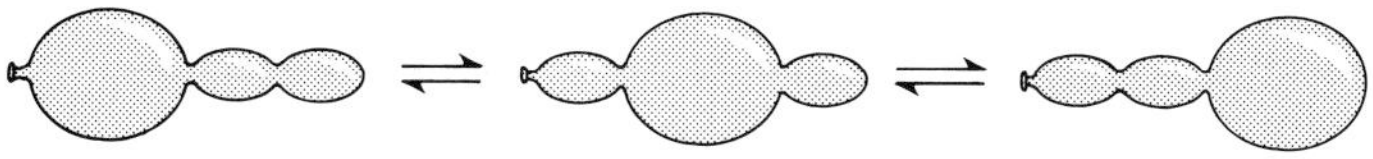

4. reactant **5. 'intermediate'** **6. product**

will have noticed that the reaction happens in two stages with the air-in-the-middle (5) being an *intermediate* stage. Each step in squeezing the air across will have a 'transition state' as before. On an energy profile it would look like this.

A catalysed reaction may involve an extra intermediate (or intermediates) in this way.

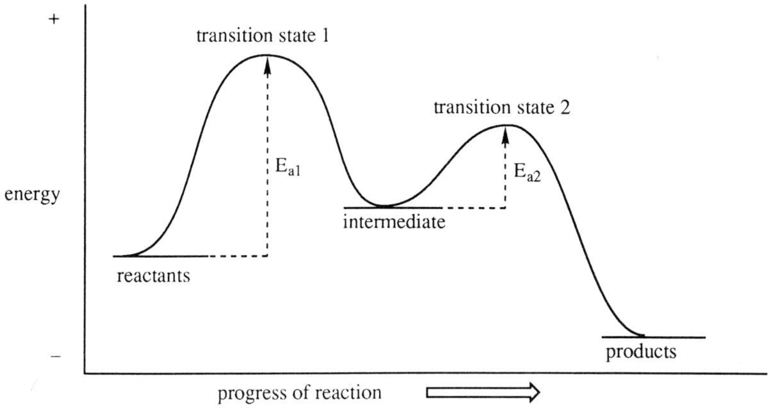

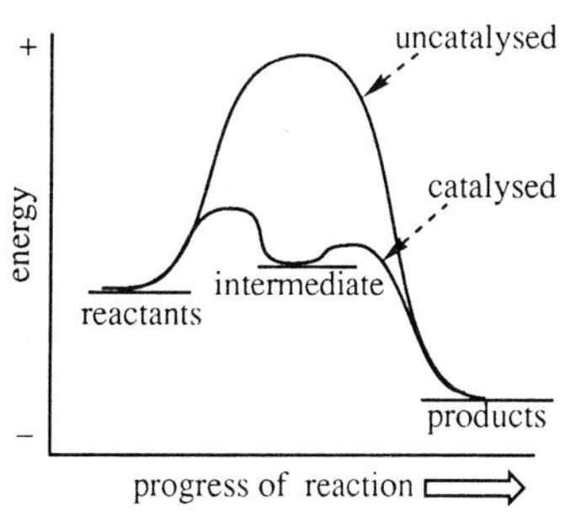

Each step has its own *activation energy* and its own *transition state*. Here the first stage is rate-determining because $E_{a1} > E_{a2}$. The *intermediate* itself has a finite existence, and can sometimes be detected or even isolated in the absence of nucleophiles such as hydroxide ions. Here is an example.

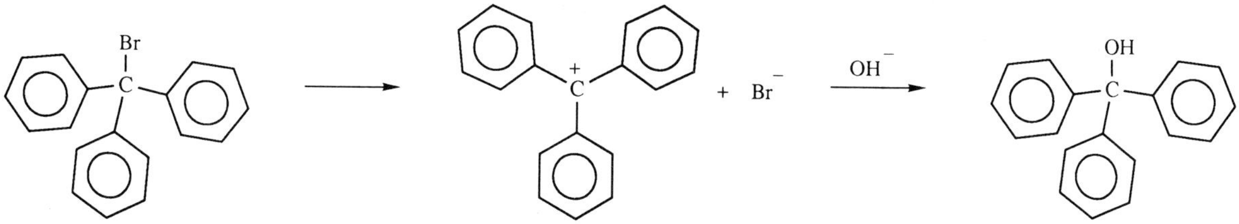

intermediate
(carbocation)

The carbocation is the *intermediate*. It has been isolated as a crystalline chlorate(VII) salt. We shall see more of these carbocations in later chapters.

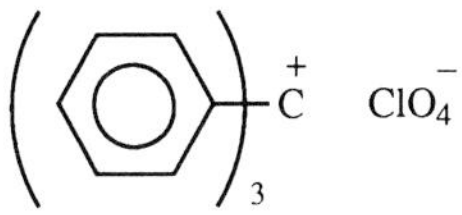

How far or how fast? Thermodynamic or kinetic control

So far in this section we have only asked two simple questions about our reactions.

1. Is the reaction likely to go on thermodynamic (or energy) grounds?
2. If the answer to the first question is yes, will the reaction go at a reasonable rate?

Now suppose that there is more than one set of possible products. The reaction of A + B can go to give either C + D or E + F as products. Which will we get and why? Let us look at the reaction profiles.

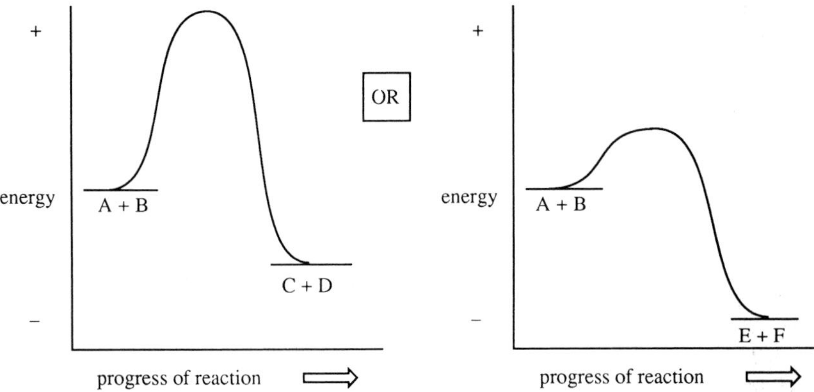

We should get the most energetically favourable products, E + F in this case, if all the reactions have reached equilibrium. This would be *thermodynamic control*. Even if the reactions were not at equilibrium the activation energy for A + B going to E + F is smaller than that for A + B going to C + D. As a result the A + B to E + F reaction is also the faster. A + B to E + F is favoured on both thermodynamic (energy) and kinetic (rate) grounds.

Now we will look at another possibility. Suppose the activation energy for A + B going to C + D were *less* than that for A + B going to E + F. What products would we get then?

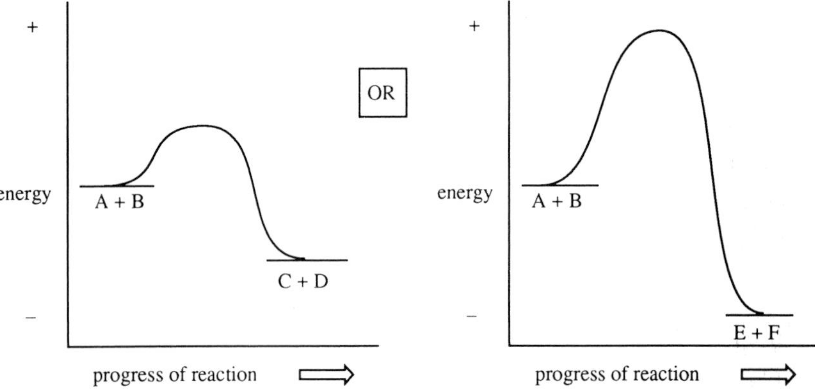

Here the thermodynamically controlled product at equilibrium will be mostly E + F, but the reaction rate to C + D will be faster (kinetic control). Keeping the temperature down will give mostly C + D, because few molecules will have enough energy for the E + F reaction. Raising the temperature gives enough energy for both reactions and their reverse; that is, they will be at equilibrium and we will get mostly E + F (thermodynamic control). An example of this is the sulphonation of naphthalene.

SO_3H

conc. H_2SO_4
80°C

conc. H_2SO_4
120°C

SO_3H

Sulponation is useful because it introduces the acidic SO_3H group, which is easily converted by neutralization into salts. e.g. $SO_3^-Na^+$.

This means that the resulting molecules are more likely to be water-soluble (see p. 15). Many detergents have long hydrocarbon tails with $-SO_3^-$ end-groups.

Which is the lower energy product? Which is produced by the faster reaction? Check your understanding by working out the probable products in the special case where the activation energies for the two possible reactions are the same. You can assume that all the reactions involved are reversible.

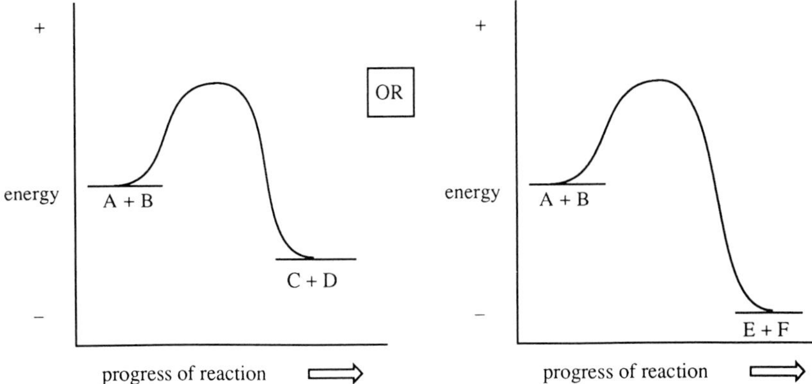

2.5 How do we get the products we want?

Chemists become quite subtle in the way they arrange for the desired outcome, even with familiar reactions.

$$CH_3CH_2OH \longrightarrow CH_3CHO \longrightarrow CH_3COOH$$

Ethanol is readily oxidized to ethanal by heating with acidified potassium chromate(VII). Ethanal is even more readily oxidized under the same conditions. The problem is how to get a good yield of ethanal. The solution is to remove it from the mixture, by distillation, as soon as it is formed. Plants and animals go one better by using, as catalysts, enzymes specific for the first reaction but not the second. These are known as alcohol dehydrogenases.

Look up the boiling points of ethanol, ethanal, and ethanoic acid, and see why this works. Rationalize the pattern in these boiling points (see p. 15).

Changing solvent can also lead to different products.

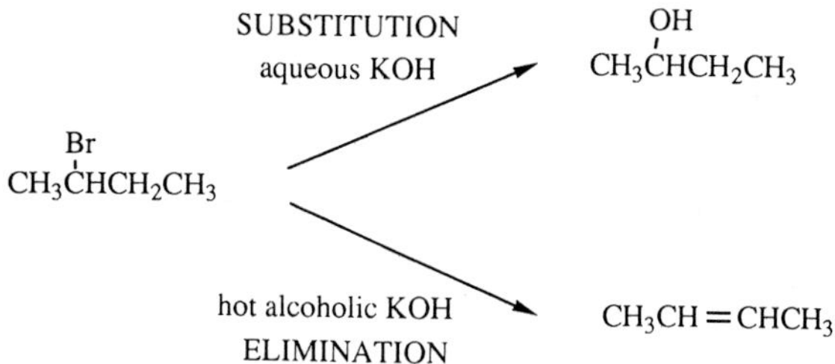

We can prepare either the alcohol or the alkene by changing the conditions. Look also at the two ways of brominating methylbenzene on p. 78.

3 Acids and bases

3.1 Introduction

The Brønsted–Lowry theory states that acids are proton donors, and bases proton acceptors. Chemists use the term 'proton' for the hydrogen ion, H^+. We will start by looking at acids and bases in water.

$$\underset{\text{acid}}{HCl_{(aq)}} + \underset{\text{base}}{H_2O_{(l)}} \rightleftharpoons H_3O^+_{(aq)} + Cl^-_{(aq)}$$

Hydrogen chloride forms ions, or dissociates, readily in water with each HCl molecule donating its proton, H^+, to water. The HCl acts as an acid, the H_2O as a base. The new O—H bond is formed using a nonbonded pair on the oxygen of H_2O; it is a dative bond.

$$H\!:\!\ddot{\underset{\cdot\cdot}{O}}\!:\;_{H} + H^+ \longrightarrow \left[H\!:\!\ddot{\underset{\cdot\cdot}{O}}\!:\!H\;_{H}\right]^+$$

Ammonia on the other hand acts as a base, whilst the water this time behaves as an acid:

$$\underset{\text{base}}{NH_{3(aq)}} + \underset{\text{acid}}{H_2O_{(l)}} \rightleftharpoons NH^+_{4(aq)} + OH^-_{(aq)}$$

We use pairs of half arrows, $\rightleftharpoons$, to represent equilibria. Acid/base reactions are largely equilibria and are therefore under thermodynamic control (see p. 29).

3.2 Equilibrium constants

A *strong* acid, such as HCl, is fully dissociated into its ions in water. The equilibrium constant for this acid dissociation is very large, and the system settles almost entirely on the right:

$$HCl + H_2O \rightleftharpoons H_3O^+ + Cl^-$$

Many organic acids, such as ethanoic acid, are *weak* acids. The equilibrium constants are small, much less than 1, and remarkably little of the acid donates its proton, H^+, to water in aqueous solution.

$$CH_3COOH_{(aq)} + H_2O_{(l)} \rightleftharpoons CH_3COO^-_{(aq)} + H_3O^+_{(aq)}$$

The equilibrium lies towards the left. The acid dissociation equilibrium constant is K_a.

$$K_a = \frac{[CH_3COO^-_{(aq)}]\,[H_3O^+_{(aq)}]}{[CH_3COOH_{(aq)}]}$$

$$= 1.7 \times 10^{-5}\ \text{mol dm}^{-3} \text{ at } 298\,\text{K}$$

$$pH = -\log_{10}[H_3O^+{}_{(aq)}]$$

$$CH_3COOH_{(aq)} + H_2O_{(l)}$$
no. of moles: $0.1 - x$

$$\rightleftharpoons$$

$$CH_3COO^-_{(aq)} + H_3O^+_{(aq)}$$
no. of moles: x x

$$K_a = \frac{[CH_3COO^-_{(aq)}][H_3O^+_{(aq)}]}{[CH_3COOH_{(aq)}]}$$

$$= \frac{x^2}{0.1 - x} = 1.7 \times 10^{-5}.$$

Because x is very small, we can put

$$\frac{x^2}{0.1} = 1.7 \times 10^{-5}.$$

Thus $x = 1.3 \times 10^{-3}$ mol dm^{-3}.

The calculation shows that, in ethanoic acid (0.1 mol dm^{-3}), the hydrogen ion concentration is only 1.3×10^{-3} mol dm^{-3}. This corresponds to a pH of 2.9. Hydrochloric acid of the same concentration has a pH of 1.0.

Most organic bases, like ammonia itself, are *weak* bases. We can write equations, similar to those for acids, to show how they act as bases in water.

$$C_2H_5NH_{2(aq)} + H_2O_{(l)} \rightleftharpoons C_2H_5\overset{+}{N}H_{3(aq)} + OH^-_{(aq)}$$

K_b is the equilibrium constant for base action in water.

$$K_b = \frac{[C_2H_5NH_3{}^+{}_{(aq)}][OH^-{}_{(aq)}]}{[C_2H_5NH_{2(aq)}]}$$

It is often more convenient to look at bases from the other end of the equation, that is to consider the acid dissociation of their *conjugate acids*, the protonated forms.

$$\underset{\text{conjugate acid}}{C_2H_5\overset{+}{N}H_{3(aq)}} + H_2O_{(l)} \rightleftharpoons C_2H_5NH_{2(aq)} + H_3O^+_{(aq)}$$

Now use K_a for the conjugate acid, $C_2H_5NH_3{}^+$.

$$K_a = \frac{[C_2H_5NH_{2(aq)}][H_3O^+{}_{(aq)}]}{[C_2H_5NH_3{}^+{}_{(aq)}]}$$

The relationship between K_a and K_b for any conjugate acid/base pair is surprisingly simple. Multiply the expressions for K_a and K_b, above, together and cancel to arrive at

$$K_aK_b = [H_3O^+{}_{(aq)}][OH^-{}_{(aq)}]$$

$$= K_w, \text{ the ionic product of water}$$

K_w has a value of 10^{-14} mol^2 dm^{-6} at 298 K.
For convenience K_a and K_b are often used in the log form:

$$pK_a = -\log_{10}K_a$$
$$pK_b = -\log_{10}K_b$$

For a conjugate acid/base pair at 298 K, $pK_a + pK_b = 14$. Large values of either pK_a or pK_b correspond to weakness.

There is a useful consequence of this relationship. The conjugate base of a strong acid will be a weak base.

$$HCl_{(aq)} + H_2O_{(l)} \rightleftharpoons H_3O^+_{(aq)} + Cl^-_{(aq)}$$

HCl is very strong acid; chloride ion, Cl^-, is a very weak base.

$$HCN_{(aq)} + H_2O_{(l)} \rightleftharpoons H_3O^+_{(aq)} + CN^-_{(aq)}$$

HCN is a fairly weak acid ($K_a = 4.9 \times 10^{-10}$ mol dm^{-3}). This makes cyanide ion, CN^-, a moderate base ($K_b = 2.0 \times 10^{-5}$ mol dm^{-3}). Ethanol is a very weak acid ($K_a = 10^{-18}$ mol dm^{-3}), so ethoxide ion is strongly basic.

$$C_2H_5OH_{(aq)} + H_2O_{(l)} \rightleftharpoons C_2H_5O^-_{(aq)} + H_3O^+_{(aq)}$$

3.3 Solubility

Many organic acids and bases are largely insoluble in water. Benzoic acid, for example, is only slightly soluble in cold water. It is also a weak acid, which means that only a small proportion of the dissolved benzoic acid will ionize as an acid.

$$C_6H_5COOH_{(aq)} + H_2O_{(l)} \rightleftharpoons C_6H_5COO^-_{(aq)} + H_3O^+_{(aq)}$$

Addition of an alkali, like NaOH, will remove H_3O^+ from the system by neutralizing it to give water. The system will shift to the right to replace the H_3O^+ and more undissolved benzoic acid will be able to go into solution. Very shortly all the benzoic acid will dissolve in excess alkali to give sodium benzoate in solution.

You might like to work through the similar argument which shows how phenylamine, $C_6H_5NH_2$, although relatively insoluble in water, dissolves freely in hydrochloric acid.

SOLUBLE ASPIRIN is the sodium salt of aspirin (see p. 1).

Why is this more soluble than aspirin itself?

$OCOCH_3$

COO^- Na^+

3.4 Reactivity of bases as leaving groups and nucleophiles

We can now try to use the acidity of the acid HX to get an idea of the ability of X^- as a *leaving group* in nucleophilic substitution, since both equations have X^- on the right hand side.

$$HX + H_2O \rightleftharpoons H_3O^+ + X^-$$

$$Nu^{:-} + \;C{-}X \longrightarrow Nu{-}C + :X^-$$

Although the second is a kinetically controlled reaction whereas the first is a thermodynamically controlled equilibrium, one can provide a guide to the other. The best leaving groups are often the conjugate bases of *strong* acids. Thus Br^- and Cl^- leave readily, CH_3COO^- less readily and OH^-, the base corresponding to the very weak acid H_2O, much less readily. This neatly explains the idea of protonating alcohols, with strong acid, to make them more open to nucleophilic attack; H_2O, the base corresponding to the strong acid H_3O^+, is a much better leaving group than OH^-. (The reaction is explained in more detail on p. 44.)

Action as a base involves electron pair donation to H^+. Nucleophilic reactions may involve electron pair donation to other atoms, such as carbon. Can we relate base strength to nucleophilic reactivity? Here are some comparisons:

Base strength:
$C_2H_5O^- > OH^- > CN^- > Cl^-$
Nucleophilic reactivity:
$CN^- > C_2H_5O^- > OH^- > Cl^-$

The concepts are linked but are not the same. Nucleophilic reactivity is measured by the *rate of reaction*, whereas base strength is measured by the *equilibrium constant*, K_b.

Nitriles (e.g. CH_3CN), on reaction with alkali, do not lose CN^- but are converted to salts of the corresponding carboxylic acids (e.g. CH_3COO^- Na^+). This is because HCN is a weak acid, and CN^- a poor leaving group. Nucleophilic *addition* to C≡N is faster.

Ethers, such as $C_2H_5{-}O{-}C_2H_5$, are resistant to nucleophilic attack because $C_2H_5O^-$, as the base corresponding to the very weak acid ethanol, is an extremely poor leaving group.

You will not be surprised to find also that the best nucleophiles are often the bases corresponding to the *weak* acids. Thus OH^-, CH_3O^-, CN^-, and CH_3COO^- are all good nucleophiles.

This is a useful guide. It cannot be more than that because we are trying to compare two situations that are very different.

3.5 Acid strengths compared

Acid strengths, recorded as K_a and pK_a values in aqueous solution, are given in Table 3.1. These are found by experiment.

Calculate the value of pK_a for water (this may not be as simple as it seems!)

Table 3.1

Acid		K_a (mol dm^{-3})	pK_a
Ethanol	CH_3CH_2OH	1.3×10^{-16}	15.9
Phenol	C_6H_5OH	1.3×10^{-10}	9.9
Hydrogen cyanide	HCN	4.9×10^{-10}	9.3
Ethanoic acid	CH_3COOH	1.7×10^{-5}	4.8
Methanoic acid	HCOOH	1.6×10^{-4}	3.8
Sulphuric acid	H_2SO_4	$\sim 10^3$	-3

Deduce which of the three OH groups of ADRENALIN is the *least* acidic.

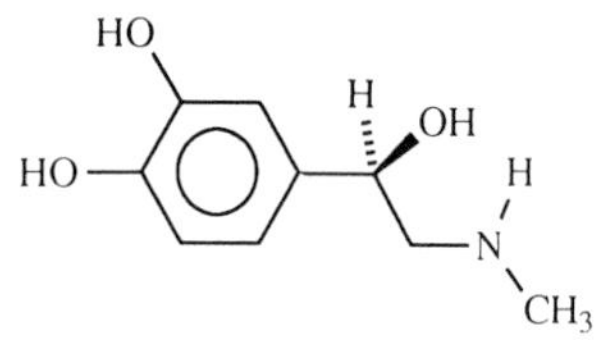

It is probably impossible to account for all the differences with a single satisfying theory. This is largely because such an account would need more data than we have available for even one case.

We can start by drawing an energy profile (see p. 23) for a generalized example of an acid dissociation.

$$HX_{(aq)} + H_2O_{(l)} \rightleftharpoons H_3O^+_{(aq)} + X^-_{(aq)}$$

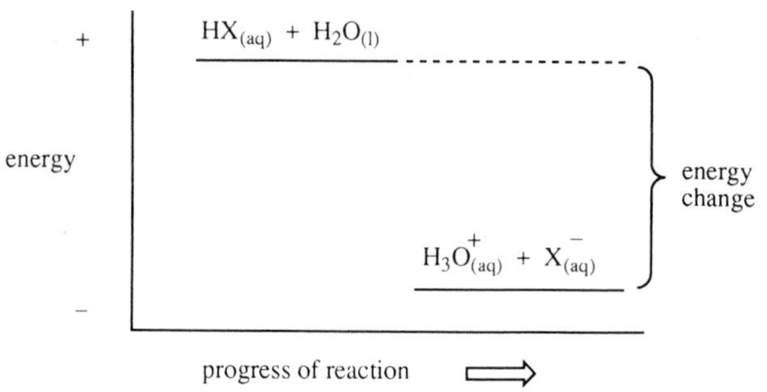

HX(aq) + H2O(l)
H3O+(aq) + X−(aq)
energy
energy change
progress of reaction

The *energy change* between reactants and products labelled in the diagram is the crucial factor. The larger, and more negative, the value of the energy difference, the higher the value of K_a, and the stronger the acid. To compare acid strengths, it is the energy differences between $HX_{(aq)}$ and $X^-_{(aq)}$ that we should be looking at. Anything which makes $X^-_{(aq)}$ more stable (that is of lower energy) will strengthen the acid. Anything which makes $HX_{(aq)}$ less stable (that is of higher energy) will have the same effect. There are three approaches we can use in trying to explain the different K_a values of organic compounds.

TENORMIN has CH, NH, and OH groups. Which is the most acidic hydrogen in the molecule? Tenormin is a drug used in the treatment of high blood pressure, angina, and abnormal heart rhythms. It acts selectively on the heart.

1. *The electronegativity of the atom carrying the negative charge in the anion X^-.* For most organic acids this is oxygen. Compounds without an electronegative atom are not usually acidic, although they can be. The pK_a for methane, for example, is 40; it is not acidic.

2. *Delocalization* of the negative charge in the anion X^- will stabilize the anion relative to the parent acid.

3. Effects of solvent, or *solvation effects*, are important but hard to quantify. HCl, for example, does not dissociate into ions much in the gas phase or in solution in methylbenzene, in spite of behaving as a very strong acid in water.

For most organic acids the energy change for dissociation in water is small and positive. This means that the equilibrium constant, K_a, is small and certainly less than 1, as can be seen in Table 3.1. They are all *weak* acids. We can now compare them and try to find reasons for their different K_a values, using the approaches listed above. As explanations they are not fundamental but they do well as guides.

Let us start with ethanol, CH_3CH_2OH, and ethanoic acid, CH_3COOH. Why should the latter be the better acid? Here the possibility of delocalization in the ethanoate anion (right) makes it reasonably stable with respect to the parent ethanoic acid. A delocalized system involves greater stability, i.e. *lower energy*, than expected from isolated single and double bonds.

Such help from delocalization is not possible for the ethanol/ethoxide system.

$CH_3CH_2-O^-$

In the same way phenol, C_6H_5OH, is a stronger acid than ethanol because of the possibility of extended delocalization in the phenoxide ion.

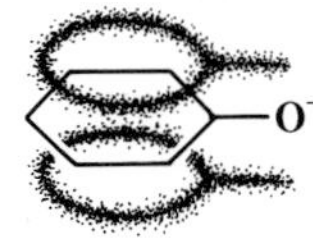

Why is phenol a weaker acid than ethanoic? Here we use the idea that delocalization is more effective for the anion which can delocalize the charge over the most oxygen atoms. Ethanoic acid wins! In the same way

$$H_2O + H_2SO_4 \rightleftharpoons HO-SO_2-O^- + H_3O^+$$

$$H_2O + H_2SO_3 \rightleftharpoons HO-SO-O^- + H_3O^+$$

sulphuric acid is stronger than sulphurous acid because there are three oxygens involved in the delocalization and not just two. Nitric and nitrous acids can be compared similarly.

3.6 Base strengths compared

PHENTERMINE is a drug used to suppress appetite. Do you think its pK_b is nearer 9 or 4?

The pK_b values for a number of organic bases are shown in Table 3.2.

Table 3.2

Base		pK_b
Phenylamine	$C_6H_5NH_2$	9.4
Ammonia	NH_3	4.8
Trimethylamine	$(CH_3)_3N$	4.2
Ethylamine	$CH_3CH_2NH_2$	3.4

The values for the various *alkyl*amines are very similar, and it seems likely that solvent effects are important in creating the small differences that are observed.

Phenylamine is notably *weak* because of the inclusion of the nonbonded pair of electrons on the nitrogen in the delocalization. Making this pair of electrons available for bonding to H^+ would involve loss of the extra delocalization energy. Therefore phenylamine is a weak base.

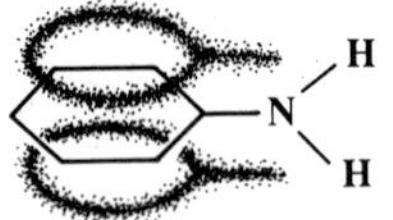

3.7 Amino acids

Amino acids are the building blocks of proteins, compounds which have major structural and catalytic roles in all living organisms.

$H_3\overset{+}{N}CH_2CO\overset{-}{O}$ glycine

$H_3\overset{+}{N}CH(CH_3)CO\overset{-}{O}$ alanine

$H_3\overset{+}{N}CH(CH_2SCH_3)CO\overset{-}{O}$ methionine

They all have the $H_3N^+CHCOO^-$ group containing both a protonated amino group and a carboxylate anion. The structures shown above are charged. Why?

Uncharged structures, such as H_2NCH_2COOH, are frequently shown in textbooks to emphasize the amino and carboxylic acid functional groups. Nevertheless they mostly exist in their charged forms, with a *very* small fraction of uncharged molecules at equilibrium.

We need to start by looking at a model for each end separately; first, CH_3COOH.

$$CH_3COOH_{(aq)} + H_2O_{(aq)} \rightleftharpoons CH_3COO^-_{(aq)} + H_3O^+_{(aq)}$$

$$K_a = \frac{[CH_3COO^-_{(aq)}]\,[H_3O^+_{(aq)}]}{[CH_3COOH_{(aq)}]}$$

From the equation it is clear that for a given weak acid the ratio $[CH_3COO^-_{(aq)}]/[CH_3COOH_{(aq)}]$ is very sensitive to the hydrogen ion concentration, or pH.

At a pH value exactly equal to the pK_a, 4.78 for ethanoic acid, the base/acid ratio will be 1. If the pH drops by 2 units, caused by added acid, 99% will be in the $-COOH$ form. Likewise, if the pH rises by 2 units above the pK_a, to 6.78 in this case, 99% will be in the $-COO^-$ form. So at pH 7 we can write it as $-COO^-$.

The nitrogen end can be given the same treatment and you should work through it for methylamine, CH_3NH_2. It will be easier to consider it starting from the conjugate acid, $CH_3NH_3^+$, whose pK_a value is 10.6. You will discover that at pH values below 8.5, the compound is mostly in its protonated form, $CH_3NH_3^+$. This is therefore the predominant form at pH 7.

Now look at the amino acids shown on p. 38. If we start, theoretically, with one $-NH_2$ and one $-COOH$ group as in H_2N-CH_2-COOH, we can show that in aqueous solution around pH 7.0 they will exist as the doubly charged structures, e.g. $H_3N^+-CH_2-COO^-$, as shown. This is also the form that they take in the crystalline state.

We can now follow what happens to glycine, for example, as we change the pH of its solution:

$H_3N^+CH_2COOH$	$H_3N^+CH_2COO^-$	$H_2NCH_2COO^-$
low pH acidic solution	pH 7 neutral	high pH alkaline solution.

The structure H_2NCH_2COOH is never the major species in aqueous solution.

Several amino acids have acidic or basic side-chains which are also involved in acid/base equilibria as the pH is changed, e.g:

aspartate

$\overset{+}{H_3N}-CH(CH_2COOH)-COOH$	$\overset{+}{H_3N}-CH(CH_2COO^-)-COO^-$	$H_2N-CH(CH_2COO^-)-COO^-$
low pH ⟸	pH 7 ⟹	high pH

You should try drawing similar diagrams for the three amino acids in the margin.

At physiological pH, around 7, any free acid or amino ends in a protein will be charged. These charges are vital in maintaining the three dimensional structure of the protein. They can also be important in binding the substrate to the enzyme, and in the catalytic action which follows. Quite small changes in pH can alter the distribution of charge in the protein, usually disastrously. Most of your enzymes only work well over a narrow pH range, e.g. around pH 8 in the gut and mouth and around pH 2 in the stomach.

This argument can be demonstrated easily using the log form of the equilibrium expression, known as the Henderson equation.

$$pH = pK_a + \log_{10}\frac{[CH_3COO^-_{(aq)}]}{[CH_3COOH_{(aq)}]}$$

At a pH 2 units below the pK_a we have

$$pK_a - 2 = pK_a + \log_{10}\frac{[CH_3COO^-_{(aq)}]}{[CH_3COOH_{(aq)}]}$$

$$\text{Thus } \log_{10}\frac{[CH_3COO^-_{(aq)}]}{[CH_3COOH_{(aq)}]} = -2$$

$$\text{and so } \frac{[CH_3COO^-_{(aq)}]}{[CH_3COOH_{(aq)}]} = \frac{1}{100}$$

You should work through the calculations for pH 6.78 in the same way.

$\overset{+}{H_3N}-CH((CH_2)_4NH_3^+)-COO^-$

lysine

$\overset{+}{H_3N}-CH(CH(CH_3)_2)-COO^-$

valine

$\overset{+}{H_3N}-CH(CH_2C_6H_4OH)-COO^-$

tyrosine

4 Reactions with nucleophiles

4.1 Introduction: nucleophiles

Nucleophiles are 'electron rich' and have either nonbonded pairs of electrons or π bonds. They can be anions or neutral molecules. Examples are

NUCLEOPHILIC ANIONS: $[\mathrm{H\ddot{\underset{\cdot\cdot}{O}}:}]^-$ usually written as HO^- or OH^-

$[\mathrm{:N{\equiv}C:}]^-$ usually written as NC^- or CN^-

NUCLEOPHILIC NEUTRAL MOLECULES: $H_2\ddot{\underset{\cdot\cdot}{O}}$, $CH_3\ddot{\underset{\cdot\cdot}{O}}H$, $\ddot{N}H_3$, $CH_3CH_2\ddot{N}H_2$.

$H_2C{=}CH_2$, $CH_3CH{=}CH_2$.

Nucleophiles can react with species which have either positive charge or low electron density. This is the basis of many reactions, which begin with the transfer of electron density from the more electron-rich atom (in the nucleophile) to the more electron-deficient atom (in the electrophile). An example is the reaction of ammonia with bromoethane.

$$H_3N\!: \; + \; H_3C{-}CH_2{-}Br \longrightarrow H_3\overset{+}{N}{-}CH_2{-}CH_3 \;\; Br^-$$

Nucleophiles and bases

Nucleophiles are also *bases* (see pp. 33 and 35) because they react with protons, H^+. Ammonia can also act as a base:

$$H_3N\!: \; + \; HCl \rightleftharpoons \overset{+}{N}H_4 \; Cl^-$$

Basicity and nucleophilicity are linked but are not the same.

Nucleophiles need electrophiles for reaction

A new bond can be formed as a nucleophilic reagent approaches an electrophile. Many organic molecules contain electrophilic carbon atoms, which are positively polarized because they are bonded to a more electronegative atom such as Cl, Br, or O (see p. 6). The electrons in the C—X bond are not evenly distributed between the two atoms and the electronegative atom readily bears a full or partial negative charge, δ^-. Examples are the haloalkanes and carbonyl compounds. Evidence for this polarization comes from the high boiling points of these compounds compared with hydrocarbons of similar size and shape (see p. 15).

We shall divide these δ^+ carbon electrophiles into three groups and then look at their reactions with nucleophiles.

Group A. Haloalkanes, e.g. CH_3CH_2Br, CH_3I.

Group B. Aldehydes and ketones, e.g. CH_3CHO (H_3C–C(=O)–H), $(CH_3)_2C{=}O$ (H_3C–C(=O)–CH_3)

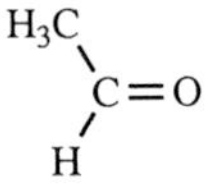

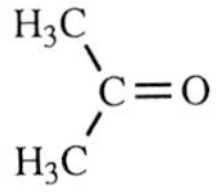

Group C. Esters, carboxylic acids and their derivatives, e.g.

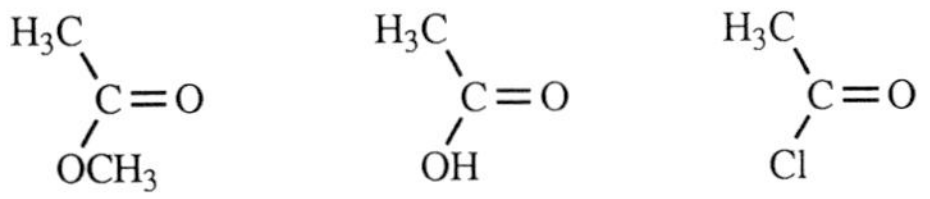

H_3C–C(=O)–OCH_3 H_3C–C(=O)–OH H_3C–C(=O)–Cl

4.2 Nucleophilic substitution reactions of haloalkanes

Bromoethane and hydroxide ion

When bromoethane is warmed with aqueous KOH, ethanol is produced. This is a *nucleophilic substitution reaction*.

The nucleophilic hydroxide ion approaches the δ^+ carbon atom of the C—Br bond. The new HO—C bond forms at the same time as the old C—Br breaks. We can draw this mechanism using dot diagrams to show how a pair of electrons on the oxygen atom forms the new bond to carbon, while the bromine atom goes off (as bromide ion) with the C—Br bond pair.

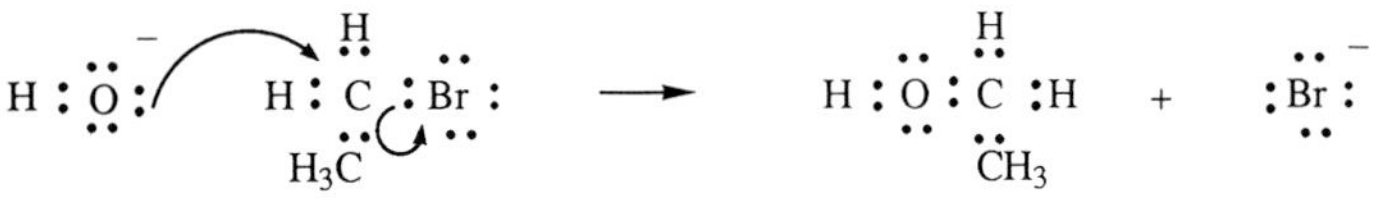

H : Ö :⁻ + H : C(H)(H₃C) : Br : ⟶ H : O : C(H)(CH₃) : H + : Br :⁻

In this mechanism, the hydroxide ion is written as HO^- so that the non-bonded pair of electrons on oxygen is placed near the δ^+ carbon of the haloalkane.

$$HO^- + CH_3CH_2Br \longrightarrow HOCH_2CH_3 + Br^-$$

This is usually abbreviated to a 'curly arrow' diagram (see p. 20).

HO^- + $H_2C(H_3C)$–Br ⟶ HO–CH_2–CH_3 + Br^-

The first arrow shows a pair of electrons from the oxygen making a covalent bond to carbon. The carbon already has eight electrons (a pair in each of four bonds) in its outer shell; so if a new pair is brought in, a pair must be lost *at the same time*. The pair the carbon loses to the Br is shown by the second arrow.

Remember that curly arrows *start at a pair of electrons:* a nonbonded pair for OH^- and the C—Br bond pair in bromoethane. We are dealing with electron pairs so these are *heterolytic* reactions (see p. 18).

This is a single-stage reaction going *via* a transition state (TS) (see p. 28).

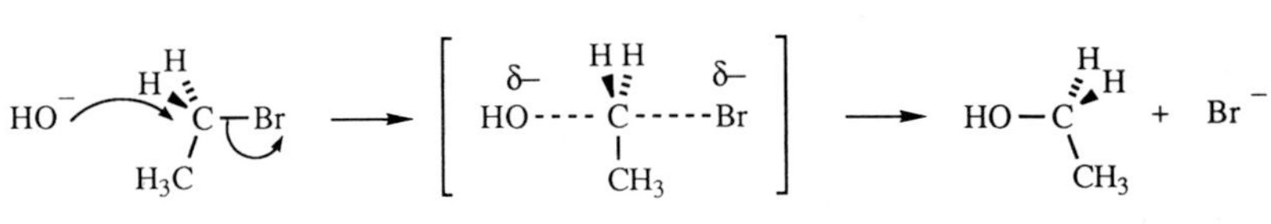

HO^- + $H_2C(H_3C)$–Br ⟶ [$\delta-$ HO---- C(H)(H)(CH₃) -----Br $\delta-$] ⟶ HO–CH_2–CH_3 + Br^-

TS

Overall this TS must have ONE WHOLE negative charge, because it is made from two species which together have one negative charge.

The shape of the transition state TS is a trigonal bipyramid (see p. 12). The transition state TS is unstable and is at an energy *maximum*.

A reaction is more likely to go if its transition state is of relatively low energy.

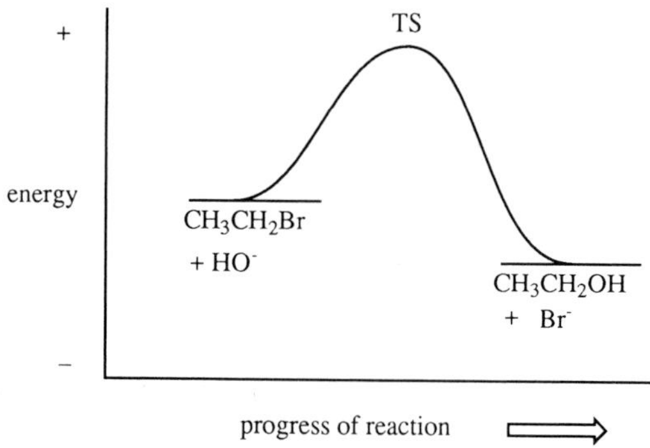

Energy profile for CH_3CH_2Br + KOH going to CH_3CH_2OH + KBr

General mechanism

Alkoxide and phenoxide ions also react with haloalkanes.
Draw the mechanism for the formation of 2,4-D

$Cl_2C_6H_3–O–CH_2–COOH$ (2,4-D)

from

$HO–C_6H_3Cl_2$ (2,4-dichlorophenol) and $Cl–CH_2–COOH$

in the presence of NaOH. 2,4-D is a selective weedkiller for broadleaved weeds in grassland.

All the reactions in this and the following five sections follow similar mechanisms. If we write Nu: or Nu:$^-$ for the nucleophile and X for the leaving group, the general mechanisms are:

$$\text{Nu:} + \text{C–X} \longrightarrow \overset{+}{\text{Nu}}\text{–C} + \text{:X}^-$$

or

$$\text{Nu:}^- + \text{C–X} \longrightarrow \text{Nu–C} + \text{:X}^-$$

The nucleophile's nonbonded pair forms the new Nu—C bond *at the same time* as the leaving group (X) goes off with the C—X bonding pair. We can now apply this general mechanism to many similar reactions.

Other haloalkanes with hydroxide ion or water

Many other haloalkanes react with aqueous hydroxide ion to give alcohols in the same way as bromoethane. For example, iodomethane gives methanol and 2-chloropropane gives propan–2–ol.

$$HO^- + (H_3C)_2CH–Cl \longrightarrow HO–CH(CH_3)_2 + Cl^-$$

Overall: $$NaOH + (CH_3)_2CHCl \longrightarrow (CH_3)_2CHOH + NaCl$$

The reactions of these haloalkanes with pure water are much slower than their reactions with aqueous NaOH or KOH. Water with its non-bonded pairs of electrons on oxygen is also a nucleophile, but a weaker one than the negatively charged hydroxyl ion (see p. 35).

Cyanide ion, CN^-

Cyanide ion reacts with haloalkanes in the same way as OH^-:

$$NC^- + H(H)(H_3C)C{-}Br \longrightarrow NC{-}C(H)(H)(CH_3) + Br^-$$

Here again the cyanide ion is drawn the other way round as NC^- to put the nonbonded pair on the carbon atom of the cyanide ion near the $\delta+$ C of the haloalkane.

Overall: $KCN + CH_3CH_2Br \longrightarrow CH_3CH_2CN + KBr$

A new C—C bond is made in the product, which is a *nitrile*. These nitriles are particularly interesting because they can be elaborated into other organic compounds. An example using iodomethane is shown below.

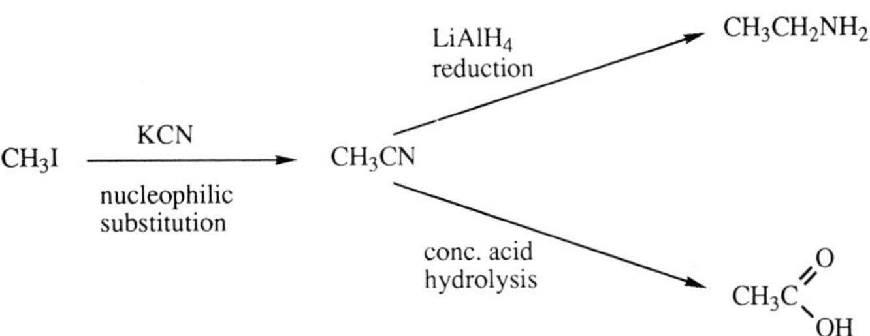

In an organic hydrolysis reaction, water is involved both as reagent and as solvent; the organic molecule is split.

In this way complex molecules can be built up.

Ammonia and the amines

Ammonia, NH_3, and the amines, such as ethylamine, $CH_3CH_2NH_2$, are nucleophiles because of the nonbonded pair of electrons on nitrogen. They react with haloalkanes by nucleophilic substitution reactions to give *salts*: e.g.

$$CH_3CH_2{-}\ddot{N}H_2 + H(H)(H_3C)C{-}Br \longrightarrow CH_3CH_2{-}\overset{+}{N}H_2{-}C(H)(H)(CH_3)\ Br^-$$

Overall: $CH_3CH_2NH_2 + CH_3CH_2Br \longrightarrow CH_3CH_2\overset{+}{N}H_2CH_2CH_3\ Br^-$

(compare $NH_3 + HCl \longrightarrow \overset{+}{N}H_4\ Cl^-$)

If you want the product *amine* and not the salt, you can *either* treat the product with a strong base to liberate the weaker base (the amine) from its salt

$$CH_3CH_2\overset{+}{N}H_2CH_2CH_3\ Br^- + KOH \longrightarrow CH_3CH_2NHCH_2CH_3 + KBr + H_2O$$

or use excess ethylamine.

$$2CH_3CH_2NH_2 + CH_3CH_2Br \longrightarrow CH_3CH_2NHCH_2CH_3 + CH_3CH_2\overset{+}{N}H_3\ Br^-$$

The amine product is still a nucleophile, and can react again with bromoethane. This can lead to mixtures of products which are difficult to separate. (Write the equations for these reactions.)

Draw the mechanism for the preparation of GLYPHOSATE

$HOOC{-}CH_2{-}NH{-}CH_2{-}P(=O)(OH)_2$

from $HOOC{-}CH_2{-}NH_2$

and $Cl{-}CH_2{-}P(=O)(OH)_2$

Glyphosate is used to control couch grass in cereals.

The natural aminoacid, glycine, can be made by a nucleophilic substitution reaction of ammonia with chloroethanoic acid. The mechanism is essentially the same as that of ethylamine with bromoethane.

$$H_3N: + ClCH_2COO^- \longrightarrow H_3\overset{+}{N}{-}CH_2COO^- + Cl^-$$

Chloroethanoate ion

Glycine (aminoethanoic acid)

4.3 Changing the leaving group: substitution reactions of alcohols

So far we have used several different nucleophiles (OH^-, H_2O, CN^-, NH_3 and amines) to react with the haloalkanes, in which the leaving group displaced is a halide ion (Cl^-, Br^- or I^-). These are good leaving groups, but for the alcohols OH^- is a poor leaving group (see p. 35). So, if we want to do nucleophilic substitution reactions on alcohols, we will need to go through a *reactive intermediate* which has a better leaving group than OH^-, such as water, H_2O. We can do this by protonating the $-OH$ of the alcohol to make $-OH_2^+$. Alcohols can be made into bromoalkanes by reaction with HBr, which is made from a mixture of KBr and concentrated H_2SO_4.

In a solution acidic enough to protonate the leaving group, many nucleophiles will themselves be protonated. For example, ammonia would be converted into the ammonium ion, which has no nonbonded pair of electrons and is not a nucleophile.

$$NH_3 + \overset{+}{H} \rightleftharpoons \overset{+}{N}H_4$$

Thus an amine *cannot* be made directly from the alcohol with NH_3, even in the presence of acid.

$$HBr \rightleftharpoons \overset{+}{H} + \overset{-}{Br}$$

$$CH_3CH_2OH + H^+ \rightleftharpoons CH_3CH_2\overset{+}{O}H_2$$

$$\overset{-}{Br} + H_3CCH_2\overset{+}{O}H_2 \longrightarrow Br{-}CH_2CH_3 + H_2O$$

$$\text{Overall: } CH_3CH_2OH + HBr \longrightarrow CH_3CH_2Br + H_2O$$

Phosphorus pentachloride reacts with alcohols in a similar way *via* a reactive intermediate, $CH_3CH_2OPCl_4$.

$$CH_3CH_2OH + PCl_5 \longrightarrow \underset{\text{intermediate}}{CH_3CH_2OPCl_4} + HCl$$

$$\overset{-}{Cl} + H_3CCH_2{-}O{-}PCl_4 \longrightarrow Cl{-}CH_2CH_3 + O{=}PCl_3 + Cl^-$$

$$\text{Overall: } CH_3CH_2OH + PCl_5 \longrightarrow CH_3CH_2Cl + POCl_3 + HCl$$

We can use these reactions in longer sequences to build up complex molecules, especially as the alcohols themselves can be prepared by reduction of aldehydes, ketones, or acid derivatives.

4.4 Polymerization of cyclic ethers

In the three-membered ring cyclic ether, $H_2C{-}CH_2$ (bridged by O), the bond angles in the ring (about 60°) are smaller than the normal bond angles in an open-chain ether (about 110°). This makes the small-ring ether unstable and reactive. It can react with nucleophiles such as OH^-.

$$HO^- + H_2C{-}CH_2 \text{ (epoxide, O bridging)} \longrightarrow HO{-}H_2C{-}CH_2{-}O^-$$

But the product is also a nucleophile and can attack another molecule of the cyclic ether, and so on until a polymer is built up:

$HO{-}CH_2{-}CH_2{-}O{-}CH_2{-}CH_2{-}O{-}CH_2 \ldots$ The repeating unit is CH_2OCH_2, an ether, so the polymer is called a *polyether*. Polyethers are used for making surfactants.

This cyclic ether can also be polymerized using an acid catalyst. Draw the mechanism for this.

4.5 Two-stage nucleophilic substitution: the carbocation intermediate

In all the reactions so far, the nucleophilic substitution has taken place in a single stage and the new bond to carbon is made at the same time as the leaving group's bond is broken.

$$Nu^- + C{-}X \longrightarrow \left[\overset{\delta-}{Nu}\cdots C \cdots \overset{\delta-}{X} \right] \longrightarrow Nu{-}C + X^-$$

transition state

This is one of the commonest mechanisms for nucleophilic substitution. However, for some compounds there is a lower energy, two-stage pathway in which the leaving group goes first and the nucleophile attacks the carbon afterwards. A positively charged, planar intermediate is formed, called a *carbocation*.

First stage: ionization

$$C{-}X \longrightarrow C^+ + X^-$$

planar, trigonal carbocation

Second stage: nucleophile reacts with cation from either side

$$Nu^- + C^+ \text{ or } C^+ + Nu^- \longrightarrow Nu{-}C \text{ or } C{-}Nu$$

Compare this two-stage reaction profile with the single-stage one (p. 42).

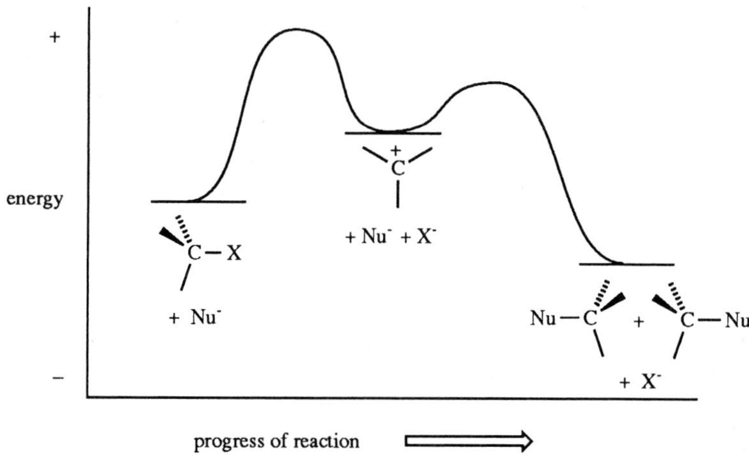

CH$_3$

C$^+$

H$_3$C CH$_3$

You can guess that the halocompounds which react by this two-stage mechanism will be the ones which can ionize to give low-energy, stable carbocations. One example is $(CH_3)_3CBr$, a crowded halide which gives a cation that is stabilized by inductive effects. A second example is $(C_6H_5)_3CBr$, which is also crowded and whose cation can be stabilized by the benzene rings.

$$(C_6H_5)_3CBr \longrightarrow (C_6H_5)_3C^+ + Br^-$$

The delocalization in each benzene ring has been extended to include the positive charge on the central C atom. This increased delocalization stabilizes the ion.

$$HO^- + (C_6H_5)_3C^+ \longrightarrow HO-C(C_6H_5)_3$$

Notice that in $(C_6H_5)_3CBr$ the halogen atom is *not* attached directly to a benzene ring, but is on the carbon *next to* the ring. Compounds such as chlorobenzene and bromobenzene, which have the halogen atom directly attached to the benzene ring, are very *unreactive* towards nucleophiles in marked contrast to the other halocompounds we have seen so far.

4.6 Competition between nucleophilic substitution and elimination

All our nucleophiles so far have reacted with electrophilic carbon atoms in the organic molecules. Nucleophiles may also act as bases and remove a proton from the haloalkane, causing *elimination* of hydrogen halide instead of substitution. The product will then be an alkene. Hence two different products can be obtained from the same haloalkane, by substitution or elimination. For example, with 2-bromopropane and KOH, *substitution* gives the alcohol (OH^- behaves as a *nucleophile*):

$HO^- + (CH_3)_2CHBr \longrightarrow HO{-}CH(CH_3)_2 + Br^-$

Overall: $KOH + (CH_3)_2CHBr \longrightarrow (CH_3)_2CHOH + KBr$

and *elimination* gives the alkene (OH^- behaves as a *base*):

$HO^- + CH_3CHBr{-}CH_3 \longrightarrow H_2O + CH_3CH{=}CH_2$

Overall: $KOH + (CH_3)_2CHBr \longrightarrow CH_3CH{=}CH_2 + H_2O + KBr$

In the elimination, the C—H bond electrons go to form the new C=C double bond. Elimination is more important if the haloalkane is very crowded (e.g. $(CH_3)_3CCl$) and if hot alcoholic KOH is used.

4.7 Reactions of nucleophiles with aldehydes and ketones

Introduction to nucleophilic addition reactions

Because the positively-polarized carbon of the carbonyl group is part of a double bond to oxygen (see Table 1.2), this time we get *nucleophilic addition* instead of the nucleophilic *substitution* of the haloalkanes. The displaced electrons of the π bond can go to the electronegative oxygen atom to make a relatively stable tetrahedral intermediate carrying a negative charge.

$Nu^- + \overset{\delta+}{C}{=}\overset{\delta-}{O} \longrightarrow Nu{-}C{-}O^-$

Tetrahedral intermediate

Draw a transition state for this reaction.

To complete the addition, this intermediate can be protonated on the $-O^-$ (see next subsection). It may then go on to eliminate water to give an 'addition–elimination' product (see p. 49).

Hydride ion donors, $NaBH_4$ and $LiAlH_4$. Aldehydes can be reduced to primary alcohols, and ketones to secondary alcohols, by either $NaBH_4$ or $LiAlH_4$. These two reducing agents can be considered as sources of nucleophilic hydride ion, H^-, which adds to the carbonyl group:

$H^- + (H_3C)_2C{=}O \longrightarrow H{-}C(CH_3)_2{-}O^-$

Tetrahedral intermediate

RASPBERRY KETONE

$HO{-}C_6H_4{-}CH_2CH_2{-}C({=}O){-}CH_3$

is used in synthetic raspberry flavourings. Draw the mechanism for the reduction of raspberry ketone by $NaBH_4$ in ethanol.

The addition is completed by protonation.

Lithium tetrahydridoaluminate, $LiAlH_4$, reacts violently with water and alcohols, so protonation is carried out after the reduction is complete.

Overall: $(CH_3)_2CO \xrightarrow[\text{then ii) } H^+_{(aq)} \text{ with great care}]{\text{i) } H^-} (CH_3)_2CHOH$

Hydrogen cyanide. HCN will also add to carbonyl compounds. This involves nucleophilic addition of CN^- then protonation by undissociated HCN or the solvent.

$\overline{N}C^- + (H)(H_3C)C{=}O \rightleftharpoons NC{-}C(H)(O^-)(CH_3)$
intermediate

$NC{-}C(H)(O^-)(CH_3) + H{-}CN \longrightarrow NC{-}C(H)(OH)(CH_3) + \overline{C}N$

Overall: $CH_3CHO + HCN \longrightarrow CH_3{-}C(H)(OH){-}CN$

Notice that the cyanide ion used in the addition is regenerated at the end; it is a catalyst and strictly it is the HCN which is used up.

The best reagents for this HCN addition are a mixture of HCN and KCN, formed by addition of cold sulphuric acid to KCN. HCN is a weak acid (see p. 33); by itself it produces a very low concentration of cyanide ions and so the reaction rate would be slow. Excess KCN is needed to increase the reaction rate by raising the cyanide ion concentration. Alternatively, a small amount of a base, such as KOH, can be added to react with HCN to increase the concentration of CN^-.

Note that these are overall *additions of HCN*. Contrast the conditions used with those for *cyanide ion substitutions* on p. 43.

Uses. The HCN addition product, with its new C—C bond, offers scope for further synthesis (see p. 43).

$(H_3C)(H)C{=}O \xrightarrow{KCN + H_2SO_4} (H_3C)(H)C(OH)(CN)$

hydrolysis $\longrightarrow (H_3C)(H)C(OH)(COOH)$

reduction $\longrightarrow (H_3C)(H)C(OH)(CH_2NH_2)$

Introduction to nucleophilic addition–elimination reactions

These reactions involve a group of nucleophiles that carry *two hydrogen atoms on a nucleophilic nitrogen atom* ($-NH_2$). This means that the usual carbonyl addition product can then eliminate water to give an unsaturated product. These are called *addition–elimination* (or *condensation*) reactions.

Hydrazine, H_2NNH_2. This is a good nucleophile. It is better than ammonia, NH_3, probably because of the repulsion between the nonbonded pairs of electrons on adjacent N atoms. One of the nucleophilic nitrogen atoms uses its nonbonded pair to add to the carbonyl group (in the same way as H^- and CN^- did). A tetrahedral intermediate (a) is formed as usual. This intermediate gains a proton on the $-O^-$ and loses a proton from the $-NH_2^+-$ group to give an uncharged addition product (b). We shall call this process, (a) to (b), a 'gain-and-loss' of protons. The uncharged intermediate (b) then eliminates water to give the final product, a *hydrazone*, which has a C=N double bond.

The structure of hydrazine is

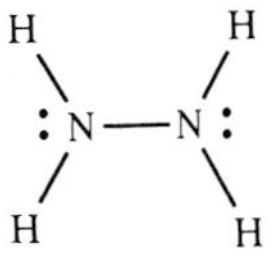

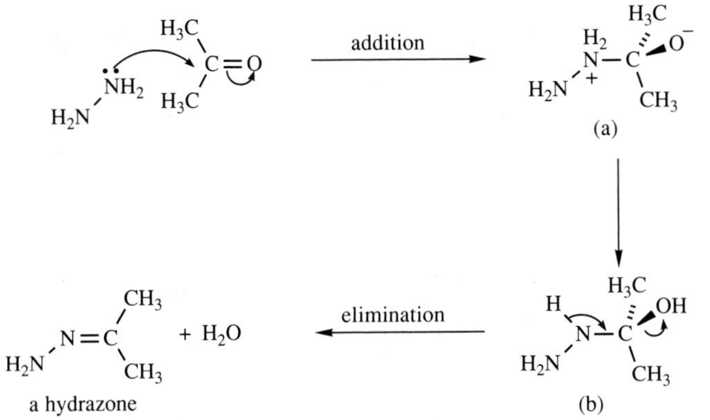

H H
:N—N:
H H

addition

(a)

elimination

+ H_2O

a hydrazone

(b)

Overall: $H_2NNH_2 + (CH_3)_2CO \longrightarrow (CH_3)_2C{=}NNH_2 + H_2O$

The final elimination stage can be catalysed by acids, which protonate the $-OH$ first. Compare this with the protonation of alcohols (p. 44).

These hydrazones are not always very easy to purify, so sometimes the 2,4-dinitrophenylhydrazones are made instead because they are easier to crystallize. The mechanism of the reaction with *2,4-dinitrophenylhydrazine* is similar to that for hydrazine.

Predict the structure of the product of the reaction of hydrazine NH_2NH_2 with LILY-OF-THE-VALLEY ALDEHYDE.

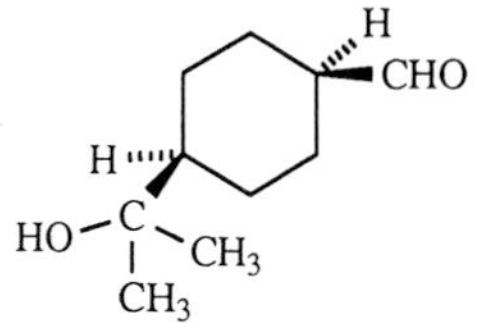

This aldehyde is used in perfumery; the *trans* compound has no detectable smell.

Overall:

+ H_2O

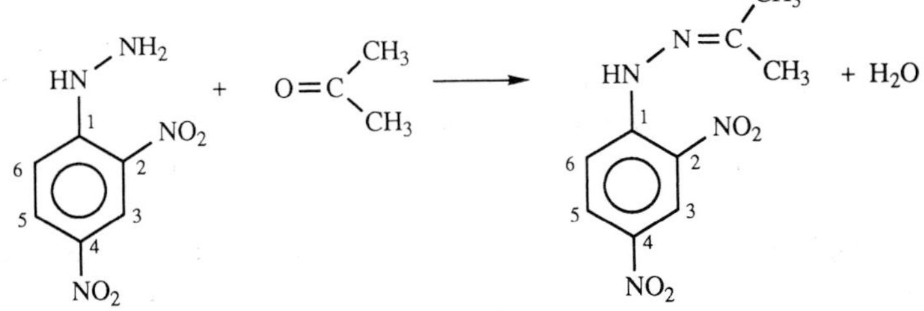

Write out the full mechanism showing the three stages:

(i) nucleophilic addition;

(ii) 'gain-and-loss' of protons;

(iii) elimination of water.

Hydroxylamine, $HONH_2$. This also reacts with aldehydes and ketones by an addition–elimination mechanism. The overall reaction for hydroxylamine and propanone is:

$$HO{-}NH_2 + O{=}C(CH_3)_2 \longrightarrow HO{-}N{=}C(CH_3)_2 + H_2O$$

Now draw out the mechanism; start by nucleophilic addition of the N nonbonded pair to the carbonyl group.

Uses. These reactions have all been used to make crystalline derivatives of aldehydes and ketones; now they are used more often in syntheses. We have exchanged a C=O for a C=NX group and made new C to N bonds. Many nitrogen-containing compounds (such as dichlobenil) can be made using this type of reaction.

2,6-dichlorobenzaldehyde (H–C=O, Cl, Cl) $\xrightarrow{NH_2OH}$ oxime (H–C=NOH, Cl, Cl)

This is part of the synthesis of DICHLOBENIL.

2,6-dichlorobenzonitrile (CN, Cl, Cl)

Draw the mechanism for this reaction. Dichlobenil is used to control weeds in apple and pear orchards; it interferes with the photosynthetic pathway.

4.8 Reactions of nucleophiles with esters, carboxylic acids, and derivatives

Introduction

Like the aldehydes and ketones, these compounds have the C=O group in their structures.

Ketones and aldehydes:

- $(H_3C)_2C{=}O$ (ketone)
- $H_3C(H)C{=}O$ (aldehyde)

Acids and derivatives:

- $H_3C(OH)C{=}O$ (carboxylic acid)
- $H_3C(OCH_3)C{=}O$ (ester)
- $H_3C(Cl)C{=}O$ (acyl chloride)

We can write a general structure CH_3COX for all of these compounds, where only X changes. The main distinction between the left- and right-hand sets is that for the right-hand set X is a possible *leaving group* (see p. 35). These can therefore undergo *addition–elimination* reactions re-forming the C=O group.

The initial stage of the reaction is the same for all: the nucleophile adds to the electron-deficient carbon of the $^{\delta+}C{=}O^{\delta-}$ to give a tetrahedral intermediate. For the acid derivatives this can now go on directly to re-form the C=O group by eliminating X (often as X^- ion): e.g.

$$\mathrm{Nu^- + H_3C(X)C{=}O \xrightarrow{\text{addition}} Nu{-}C(H_3C)(O^-)(X)\ (\text{intermediate}) \xrightarrow{\text{elimination}} CH_3(Nu)C{=}O + X^-}$$

Overall: $\mathrm{Nu^- + CH_3COX \longrightarrow CH_3CONu + X^-}$

An example is the conversion of an ethyl ester into a methyl ester (a *transesterification* reaction).

$$\mathrm{H_3CO^- + H_3C(OC_2H_5)C{=}O \rightleftharpoons H_3CO{-}C(H_3C)(O^-)(OC_2H_5) \rightleftharpoons H_3CO{-}C(=O)CH_3 + {}^-OC_2H_5}$$

The *overall* result is a substitution of the nucleophile Nu for X. There are two stages and the first addition stage is usually the slower, rate-determining stage.

Relative reactivity of different derivatives

When we compare the reactivity of the various carboxylic acid derivatives, RCOX, two factors operate in opposite directions.

1. The *inductive effect* of X. If X is more electronegative than C, this reduces the electron density on the C=O carbon atom even more, and so would *increase* the likelihood of bond formation to nucleophiles.

2. The stabilization of the molecule by *delocalization* of a lone pair of electrons on X with the C=O double bond. This will decrease the reactivity to nucleophiles, as the energy gap between the more stable starting material and the transition state, which does not have this delocalization, will probably be greater.

Inductive effect: $\mathrm{\overset{2\delta+}{C}{=}\overset{\delta-}{O}}$, $\mathrm{X^{\delta-}}$

Delocalisation: —C(O)X

Note that the observed flat structure of the amide group, X = NH_2, with its angles of about 120° around carbon and nitrogen, would be predicted from this delocalization. This flat amide or peptide structure is very important in maintaining the shapes of proteins and enzymes.

Now look at the balance of inductive effects and delocalization on the reactivity of CH_3COX.

Esters and acids also have a flat —C(=O)—O— system like the amides —C(=O)—N<

1. *Acyl halides* (also called acid halides), e.g. X = Cl (CH_3COCl). Here the inductive effect of the electronegative halogen seems to dominate and these are very reactive to nucleophiles.

2. *Esters and amides*, e.g. X = OCH_3 (CH_3COOCH_3) and X = NH_2 (CH_3CONH_2). Here delocalization reduces the reactivity; for example, $NaBH_4$ will reduce acyl halides, aldehydes, and ketones but not esters and amides.

Let's start off by looking at the most familiar reaction of esters, ester hydrolysis, and its reverse, esterification.

Esters

Hydrolysis and esterification. Esters are hydrolysed to carboxylate salts and alcohols when warmed in aqueous alkaline solution, e.g. for ethyl ethanoate:

$$CH_3COOCH_2CH_3 + NaOH \longrightarrow CH_3CO\bar{O}\ \overset{+}{Na} + CH_3CH_2OH$$

The standard *addition–elimination* mechanism outlined on p. 51 is followed, ending with a proton exchange between the acid and the alkoxide ion.

(i) Nucleophilic *addition* of $H\bar{O}$ to $\overset{\delta+}{C}=\overset{\delta-}{O}$ to give a tetrahedral intermediate A.

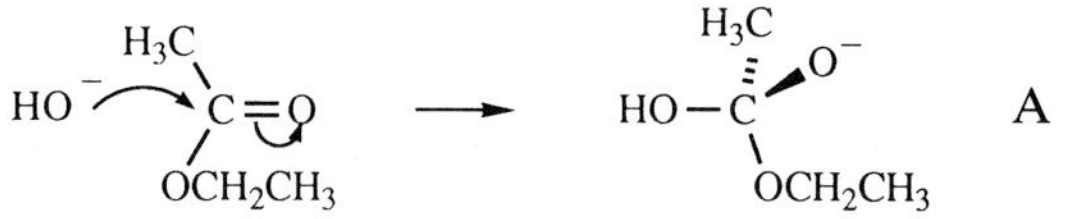

$$H\bar{O} + (H_3C)(OCH_2CH_3)C=O \longrightarrow HO-C(H_3C)(O^-)(OCH_2CH_3) \quad \text{A}$$

(ii) Intermediate A re-forms the C=O bond with *elimination* of alkoxide ion, $CH_3CH_2O^-$.

$$\text{A} \quad HO-C(H_3C)(O^-)(OCH_2CH_3) \longrightarrow HO-C(CH_3)=O + \bar{O}CH_2CH_3$$

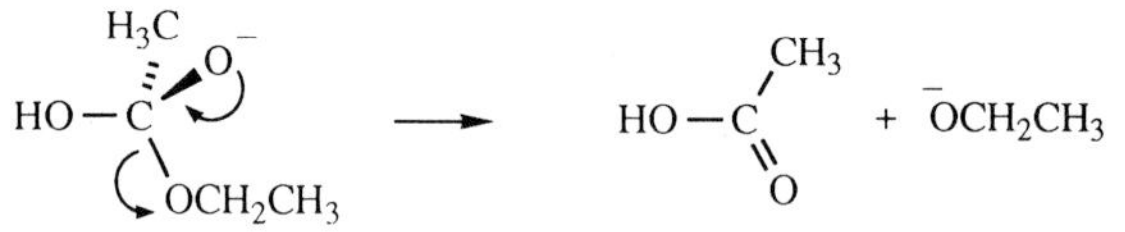

(iii) *Proton exchange* then occurs between the acid and alkoxide ion; since $CH_3CH_2O^-$ is a stronger base than CH_3COO^-, this equilibrium lies far over to the right causing the hydrolysis to be essentially irreversible.

$$CH_3CH_2\bar{O} + H-O-C(CH_3)=O \rightleftharpoons CH_3CH_2OH + \bar{O}-C(CH_3)=O$$

This is a very good way to hydrolyse esters, and it is used in the hydrolysis of natural esters (fats) for soap production.

Esters can also be hydrolysed in aqueous acidic solution. The C=O group is first protonated to make the carbonyl carbon even more electron-deficient; this is necessary because water is a weaker nucleophile than OH^-. See if you can complete the mechanism for this.

$$(H_3C)(OCH_2CH_3)C=\ddot{O}: + \overset{+}{H} \rightleftharpoons (H_3C)(OCH_2CH_3)C=\overset{+}{O}H$$

$$\text{Overall: } \overset{+}{H} + CH_3COOCH_2CH_3 + H_2O \rightleftharpoons CH_3COOH + CH_3CH_2OH + \overset{+}{H}$$

The proton is regenerated at the end so this is an *acid-catalysed* hydrolysis. Since it is a reversible reaction, an equilibrium, we can also use it to *prepare* esters. We still need the acid catalyst, in order to reach equilibrium faster, but we must now start off with as *little* water as possible so that as much ester as possible is present at equilibrium.

Thus, if we choose the conditions carefully we can use the acid-catalysed reaction for either ester hydrolysis or esterification.

PYRETHRIN I is a natural insecticide made by some chrysanthemum species.

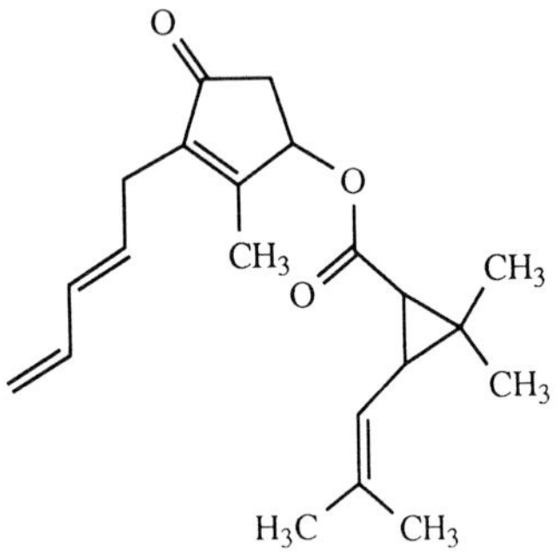

It is an ester. Draw the structures of the acid and the alcohol from which it might be synthesized.

$$CH_3COOCH_2CH_3 + H_2O \rightleftharpoons CH_3COOH + CH_3CH_2OH$$

ESTER HYDROLYSIS →
H^+ and water i.e. aqueous acid

← ESTERIFICATION
H^+, no water i.e. conc. H_2SO_4 or dry HCl

If there is *no acid or base* present, just ester and water, the reaction is so slow that nothing appears to happen at all.

The importance of these reactions can be seen from the wide range of naturally occurring esters and from the number of biological catalysts, the *esterase enzymes*, used by plants and animals for these processes.

Lithium tetrahydridoaluminate reduction of esters. This goes in several stages, the aldehyde being an intermediate. The mechanism is a combination of:

(i) typical addition–elimination of esters; with

(ii) nucleophilic addition to carbonyl (see p. 47).

We will use H^- for the $LiAlH_4$.

i) H^- + H_3C–C(=O)–OCH_2CH_3 —addition→ H_3C–C(H)(O^-)(OCH_2CH_3) —elimination→ H–C(CH_3)=O + $^-OCH_2CH_3$

ii) H^- + H_3C–C(=O)–H —addition→ H_3C–C(H)(H)–O^- —protonation, $+ H_2O$→ H_3C–C(H)(H)–OH + OH^-

Overall: $CH_3COOC_2H_5 + 4H \longrightarrow CH_3CH_2OH + C_2H_5OH$

Notice that $NaBH_4$, which is less reactive than $LiAlH_4$, does *not* normally reduce either esters or acids but *will* reduce aldehydes and ketones.

Benzene-1,4-dicarboxylic acid and ethane-1,2-diol react together to give a polyester, terylene. Draw the repeating unit of terylene.

COOH
COOH

Amides

Amides and peptides contain the same planar functional group (see p. 51). Many important natural products (proteins, enzymes, hormones, antibiotics) and synthetic polymers (nylon) contain this group. Amides and peptides are generally less reactive to nucleophiles than esters, but like esters they can be hydrolysed in either acidic or alkaline solution by nucleophilic addition–elimination mechanisms.

Overall: $CH_3CONH_2 + H_2O + H^+ \longrightarrow CH_3COOH + \overset{+}{N}H_4$

or $CH_3CONH_2 + NaOH \longrightarrow CH_3COO^- Na^+ + NH_3$

O
C
N

amide or peptide link

Amides can be made by nucleophilic reaction of ammonia or amines on acyl chlorides, acid anhydrides (see below), or esters, again by addition–elimination mechanisms. Note that ammonia (a *base*) reacts with a carboxylic *acid* to give a salt.

$$CH_3COOH + NH_3 \longrightarrow CH_3CO\overset{-}{O}\ \overset{+}{N}H_4$$

Pick out the peptide link in the drug ZESTRIL, used in the treatment of high blood pressure and heart failure.

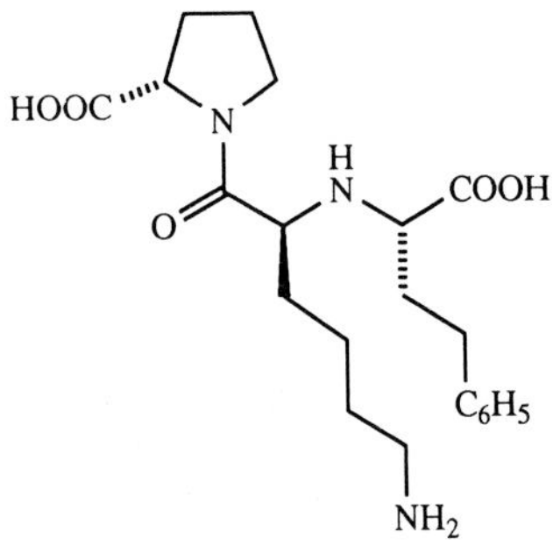

Draw the structures of the two aminoacids from which this peptide could be made.

Proteins are polymers of aminoacids, joined together by amide (peptide) bonds.

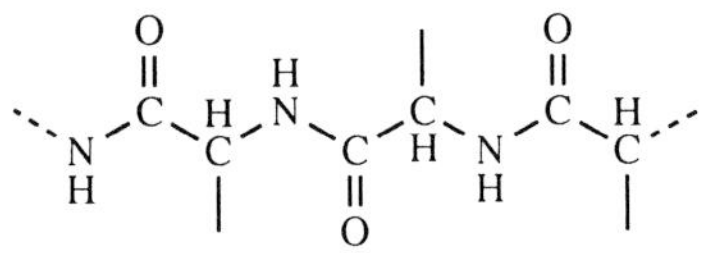

In the laboratory, these are usually hydrolysed back to a mixture of aminoacids using acidic conditions because some of the natural aminoacids are unstable in alkaline solution. In animals and plants, proteins are hydrolysed using natural catalysts, enzymes such as chymotrypsin. An interesting aspect of enzyme-catalysed hydrolysis reactions is that these, too, use the standard mechanism:

(i) nucleophilic addition to C=O to give a tetrahedral intermediate;

(ii) re-formation of the C=O group and elimination of a leaving group.

Acyl chlorides and anhydrides

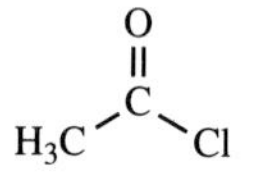

ethanoyl chloride

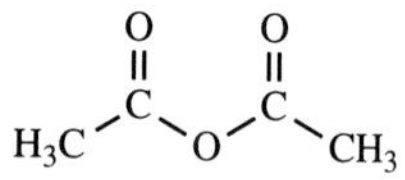

ethanoic anhydride

These all have the group RCOX where X is Cl or $OCOCH_3$. They are *more reactive* towards nucleophiles than esters. Ethanoyl chloride, CH_3COCl, reacts almost explosively with cold aqueous NaOH (*don't* do it!) and even its reaction with water is violent. Anhydrides come in between the acyl chlorides and the esters in reactivity.

Both families of compounds react readily by the standard addition–elimination mechanism with nucleophiles such as water, alcohols, ammonia, and amines. Work out the mechanisms for these two examples:

$$C_6H_5COCl + CH_3CH_2OH \longrightarrow C_6H_5COOCH_2CH_3 + HCl$$

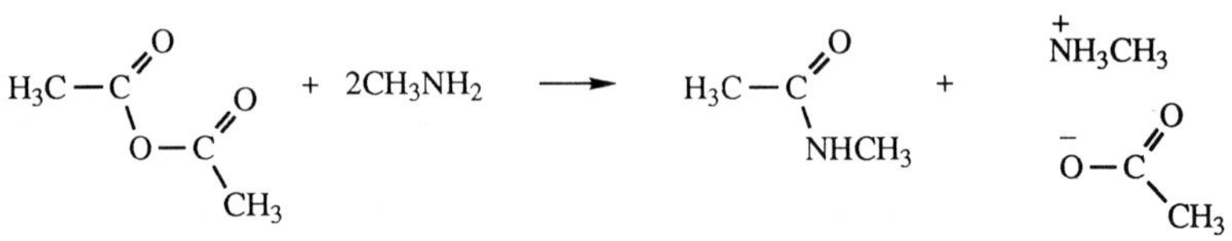

$$(CH_3CO)_2O + 2CH_3NH_2 \longrightarrow CH_3CONHCH_3 + CH_3COO^- \ \overset{+}{N}H_3CH_3$$

You should be able to see why *two* molar equivalents of amine are used in the second example.

Acid anhydrides can be made from acyl chlorides and *anhydrous* carboxylate salts. Write out this mechanism, too.

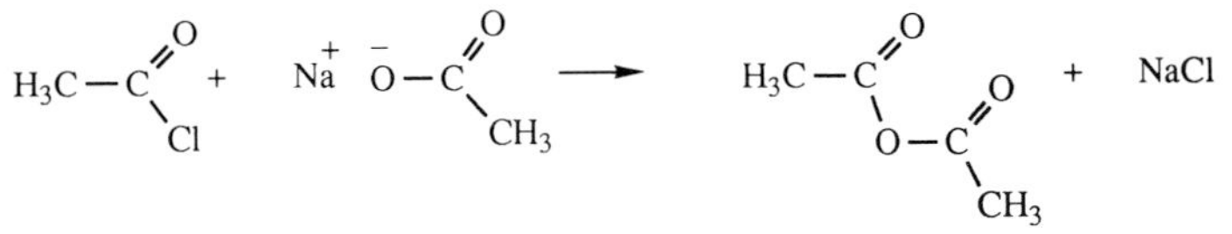

$$H_3C-C(=O)Cl + \overset{+}{Na}\ \overset{-}{O}-C(=O)CH_3 \longrightarrow H_3C-C(=O)-O-C(=O)CH_3 + NaCl$$

Overall: $CH_3COCl + CH_3CO\overset{-}{O}\ \overset{+}{Na} \longrightarrow CH_3COOCOCH_3 + NaCl$

One of the intermediates in the synthesis of the foliar fungicide CAPTAN

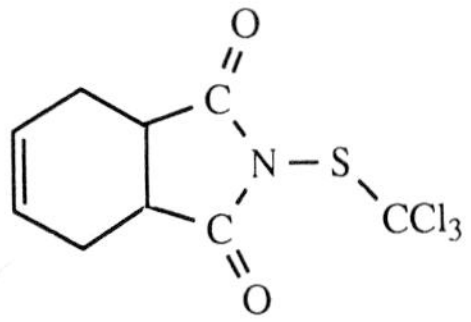

is the acid anhydride

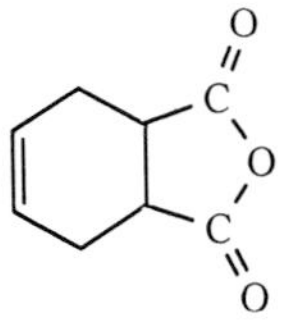

Predict the structure of the product formed when this anhydride reacts with ammonia, and write the mechanism.

Captan is used against apple and pear scab. It reacts with other compounds in the fungus to produce toxic $CSCl_2$.

Carboxylic acids

Nucleophiles are also bases and often react with carboxylic acids to form salts. For example,

$$CH_3COOH + NaOH \longrightarrow CH_3COO^{-}\ \overset{+}{Na} + H_2O$$

This limits the reactivity of carboxylic acids with nucleophiles, because the resulting carboxylate anion is much less reactive to nucleophiles than esters or acid anhydrides.

$$CH_3COOH + H_2NCH_2CH_3 \longrightarrow CH_3CO\overset{-}{O}\ H_3\overset{+}{N}CH_2CH_3$$

Under forcing conditions amides can be made from carboxylic acids, e.g. in the polymerization to give nylon-6.

$H_2N(CH_2)_5COOH$ gives ---- $NH(CH_2)_5CONH(CH_2)_5CONH(CH_2)_5CO$ ----

Carboxylic acids react with inorganic acid chlorides such as $SOCl_2$ or PCl_5 to give organic acid chlorides; compare this with the conversion of alcohols into haloalkanes (see p. 44).

$$CH_3COOH + SOCl_2 \xrightarrow{\text{dry}} CH_3COCl + SO_2 + HCl$$

The reduction of acids by $LiAlH_4$ is similar to the reduction of esters (see p. 53).

Overall: $CH_3CH_2COOH + 4H \longrightarrow CH_3CH_2CH_2OH + H_2O$

4.9 Comparison of acid derivatives with aldehydes and ketones

Aldehydes and ketones differ from the acid derivatives because they have H, alkyl, or phenyl instead of a possible leaving group, X. They all undergo addition first to give a tetrahedral intermediate. Now the best leaving group in the case of the aldehydes and ketones is the nucleophile (—Nu) and its loss simply reverses the addition. For example, for propanone with cold, dilute OH^-:

The intermediate can exchange its —OH proton with water

$$H_2O + HO{-}C(CH_3)_2O^- \rightleftharpoons HO^- + HO{-}C(CH_3)_2OH \rightleftharpoons H_2O + {}^-O{-}C(CH_3)_2OH$$

Using this, work out what happens when propanone is treated with cold, dilute NaOH in $H_2{}^{18}O$.

$$HO^- + (H_3C)_2C{=}O \longrightarrow HO{-}C(CH_3)_2O^- \text{ then } HO{-}C(CH_3)_2O^- \longrightarrow HO^- + (H_3C)_2C{=}O$$

intermediate

4.10 Comparison of the reactivities of the different types of halocompound with nucleophiles

We will now compare the reactions of three groups of chlorocompounds with aqueous OH^- and with H_2O.

Group I. Alkyl halides, including $CH_3CH_2CH_2Cl$ and $C_6H_5CH_2Cl$.

Group II. Acyl halides, such as CH_3COCl.

Group III. Phenyl and aryl halides, such as C_6H_5Cl, and haloethenes, such as $H_2C{=}CHCl$.

Group I. Here the Cl is attached to a carbon atom which carries only single bonds; notice that in Groups II and III the Cl is attached to a double-bonded carbon atom. For Group I we therefore expect (and find) direct displacement, and we look for addition and then elimination in Groups II and III.

Group II. Addition of nucleophiles to C=O is very easy; the C=O group is highly polarized and oxygen takes the pair of π electrons from C=O and so bears the negative charge in the intermediate.

$$Nu^- + H_3C(Cl)C{=}O \longrightarrow Nu{-}C(CH_3)(Cl)O^- \longrightarrow H_3C(Nu)C{=}O + Cl^-$$

Tetrahedral intermediate

These Group II reactions are very fast.

Group III. Addition of nucleophiles to C=C is normally difficult; the C=C group is 'electron-rich' and not very polarized. If addition did occur, the negative charge would be on carbon, e.g. structure B.

$$HO^- + C_6H_5Cl \xrightarrow{???} \text{B}$$

B

Presumably these intermediates are of high energy and are difficult to form, because the Group III halides are essentially inert to OH^- under normal conditions. Notice how the reactivity changes sharply according to whether the Cl is *directly* attached to a benzene ring (Group III) or attached to a carbon next to a benzene ring (Group I).

The reactivities of the three groups, with variations according to precise structure, are summarized in Table 4.1.

Table 4.1

Group	II	I	III
Type	Acyl halides	Alkyl halides	Aryl halides
Reaction with aq. OH^-	Violent, almost explosive	Fast	No reaction
Reaction with H_2O	Very fast	Slow	No reaction

5 Reactions with electrophiles

5.1 Introduction

Electrophiles have centres of low electron density, which will accept an electron pair to make a covalent bond. There are three types: positively charged cations, neutral molecules, and radicals. The first two will be dealt with here (see p. 75 for radicals).

Notice that the stable cations Na^+ and K^+ are not included because they do not easily accept extra electrons to form covalent bonds.

Cations, e.g. H^+ from HCl, NO_2^+ from HNO_3. These cations can readily accept a pair of electrons to form new covalent bonds.

Neutral molecules, e.g. Br_2, HBr. Some of these are easily polarized, such as bromine, $^{\delta+}Br{-}Br^{\delta-}$; the $\delta+$ part is electrophilic. Others are permanently polarized, e.g. hydrogen bromide (see p. 6), $^{\delta+}H{-}Br^{\delta-}$.

We could also include as electrophiles all the organic molecules which react with the nucleophilic reagents in Chapter 4, e.g. the polarized haloalkanes and carbonyl compounds. To prevent repetition, we will follow the usual convention in organic chemistry and classify the inorganic and small organic species as the reagents. This can lead us into trouble when neither reactant fits into the 'reagent' category!

Addition to double bonds

The electron-rich organic molecules in this chapter all have formal carbon–carbon double bonds. The carbon–carbon double bond can be considered as a σ bond and a π bond. The electrons of the π bond are in a π molecular orbital, made by combining two p atomic orbitals. This π orbital has its average electron density further from the positive carbon nuclei than in a single σ bond. The electrons are more easily available to electrophiles.

The carbon–carbon double bond is weaker than two single bonds and so electrophiles (generalized as E^+) normally *add* to C=X. In the addition to C=C, electrons are transferred from the double bond to E^+, creating positively charged intermediates.

$$\text{C=C} + E^+ \longrightarrow \text{E–C–C}^+$$

intermediate

A new C—E bond is formed leaving the other carbon atom as an electron-deficient *carbocation*, with only six electrons in its outer shell. This may be clearer from a dot diagram taking ethene as our example.

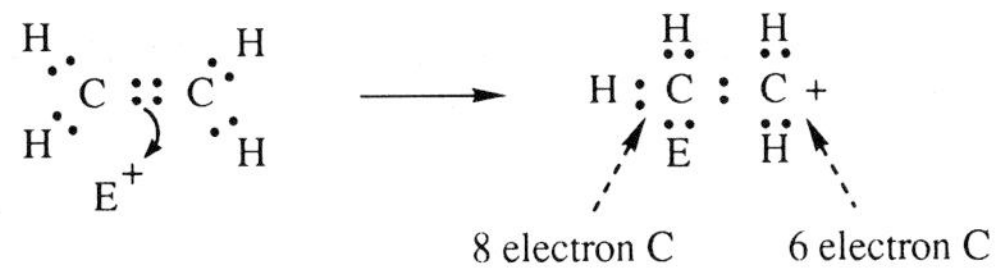

8 electron C　　6 electron C

The electrons of the π bond *both* go to form the new C—E bond, so this is a *heterolytic* fission of the bond. The reaction can then be completed by addition of an anion, X^-.

$$H_2C=CH_2 + E^+ \longrightarrow ECH_2-\overset{+}{C}H_2$$

$$ECH_2-\overset{+}{C}H_2 + X^- \longrightarrow ECH_2-CH_2X$$

$$\text{Overall:}\quad H_2C=CH_2 + EX \longrightarrow ECH_2-CH_2X$$

Both steps involve the formation of new *dative* covalent bonds.

We shall see later that compounds like benzene follow the same first addition stage, but then lose H^+ instead of gaining an anion.

5.2 Addition of hydrogen halides to alkenes

Ethene

When hydrobromic acid and ethene are mixed, oily drops of bromoethane are formed.

$$H_2C=CH_2 + HBr \longrightarrow H_3C-CH_2Br$$

The electrophile is the proton, from the dissociation of HBr, a strong acid.

$$HBr_{(aq)} \rightleftharpoons \overset{+}{H}_{(aq)} + \overset{-}{Br}_{(aq)}$$

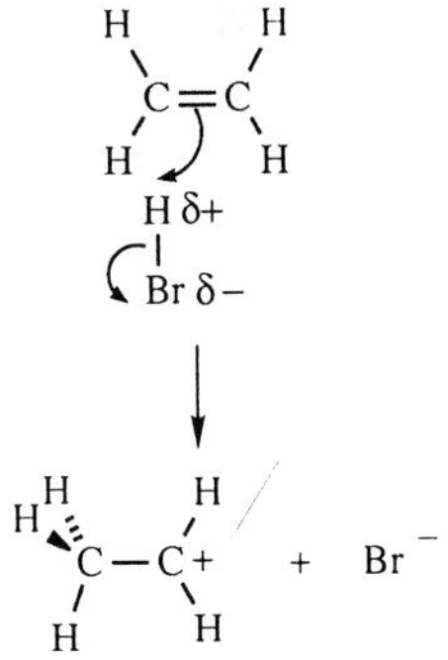

The proton adds to the carbon–carbon double bond to form a new C—H bond. The carbocation intermediate then adds the Br^- to give bromoethane, in which both C atoms now have eight outer shell electrons. For dot diagrams this is:

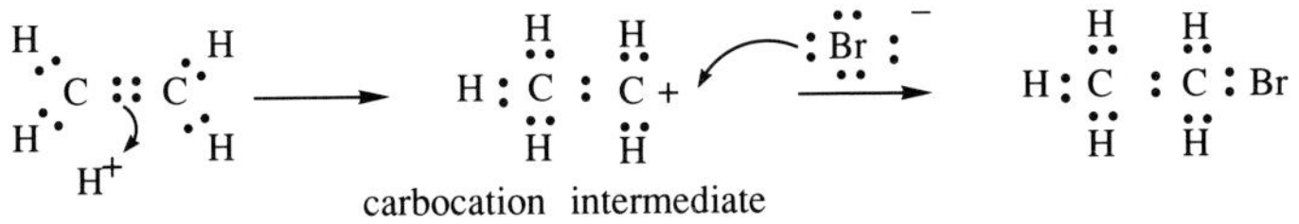

carbocation intermediate

Or, using curly arrows (remember that the arrow *begins at the electron pair* and goes to make a bond to the electron-deficient atom):

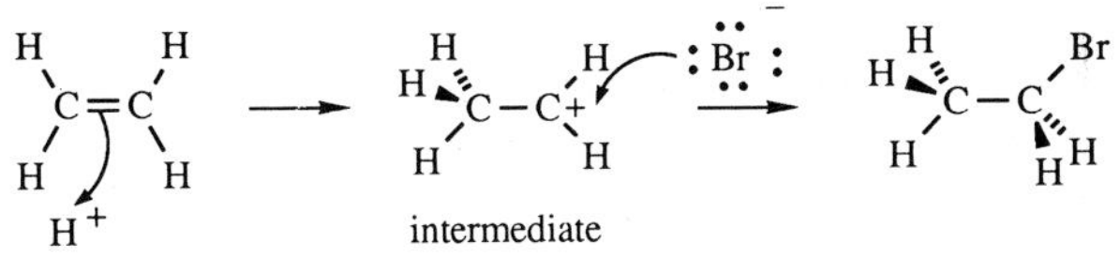

intermediate

If the HBr is not dissociated, the polarized $\overset{\delta+}{H}-\overset{\delta-}{Br}$ molecule can act as the electrophile to give the same carbocation intermediate.

The reaction of ethene with HCl follows a similar mechanism (write it out, using both types of diagram).

As the electrophile here is a proton, H^+, the first addition to C=C is also an *acid–base* reaction (see p. 33). HBr is the acid and ethene is the base. Because alkenes are weak bases only strong acids protonate them effectively, so that HCN (a weak acid) does not usually add to alkenes.

Propene

The initial addition of H^+ to propene can give two different carbocation intermediates, one primary (P) and one secondary (S).

Primary carbocations have only *one* C substituent attached to the C^+, secondary have *two*, and tertiary (e.g. $(CH_3)_3C^+$) have *three*.

Either $H_3C(H)C{=}CH_2 + H^+ \longrightarrow H_3C(H)_2C{-}C^+H_2$ (P)

or $H_3C(H)C{=}CH_2 + H^+ \longrightarrow H_3C(H)C^+{-}CH_3$ (S)

The secondary carbocation (S) is more stable (of lower energy) than the primary one (P) and is formed faster in this first, rate-determining step of the reaction. We assume that the transition state leading to the secondary carbocation is of lower energy than the corresponding one leading to the primary carbocation. Let's consider two possible reasons for this: stereochemical and inductive.

More stable (lower energy) intermediates are likely to be preceded by lower energy transition states, since the structure of a transition state is probably similar to the structure of the intermediate.

Stereochemical effects. For a stable three-coordinate carbocation, the ideal angles around C^+ are about 120° (three electron pairs, as far away from each other as possible). This keeps large substituents farther apart than the 109° (tetrahedral) bond angle of a 4-coordinate carbon. Thus the change from four- to three-coordination favours the more stereochemically crowded carbon; tertiary > secondary > primary for stability of carbocations.

stability $(H_3C)(CH_3)(CH_3)C^+$ > $(H)(CH_3)(CH_3)C^+$ > $(H)(CH_3)(H)C^+$

Inductive effects. The electron-releasing inductive effect of an extra methyl group (see p. 7) helps to neutralize and stabilize the developing positive charge on carbon. Tertiary carbocations with three methyl groups, such as $(CH_3)_3C^+$, are even more stable than secondary carbocations.

stability $(CH_3)_2\overset{+}{C}CH_3$ > $H(CH_3)\overset{+}{C}CH_3$ or $F_3C(CH_3)\overset{+}{C}CH_3$

In general, the more stable the intermediate, the lower its energy and the more likely it is to be formed. Therefore the addition of HBr to propene goes mostly *via* the more stable secondary carbocation (S) to give 2-bromopropane, rather than *via* the less stable primary carbocation (P) to give 1-bromopropane. In fact, the product is nearly all 2-bromopropane.

$$H_3C-CH=CH_2 \xrightarrow{H^+} H_3C-\overset{+}{C}H-CH_3 \ (S) \xrightarrow{Br^-} H_3C-CHBr-CH_3$$

Draw the mechanism for the addition of molecular HBr (*not* ionized) to propene.

For propene and HCl, the 2-halopropane is again the major product.

If you want to predict which haloalkane will be formed by polar addition of hydrogen halide to an unsymmetrical alkene (like propene), draw the possible carbocation intermediates. Work out which you think is the more stable; this intermediate should be formed more easily and lead to the major product.

Try this idea for the addition of HBr to 1–pentene and to $F_3CCH{=}CH_2$.

Markovnikov's rule, which states that 'the more positive Y portion of the reagent Y–Z goes to the carbon atom of the C=C which already bears the more H atoms', is an empirical rule governing these additions, and was used before the mechanisms were understood.

5.3 Reactions of alkenes with sulphuric acid: hydration

Ethene

Concentrated sulphuric acid (a strong acid) will protonate the double bond of ethene just as HCl and HBr do.

$$CH_2{=}CH_2 \xrightarrow{H_2SO_4} CH_3-\overset{+}{C}H_2 \quad HSO_4^-$$

carbocation intermediate

The HSO_4^- anion traps the carbocation to give a hydrogensulphate ester, which is easily hydrolysed by water.

$$H_2C{=}CH_2 + H-OSO_3H \longrightarrow H_3C-\overset{+}{C}H_2 \xrightarrow{HSO_4^-} H_3C-CH_2-OSO_3H$$

a hydrogensulphate ester

$$H_3C-CH_2-OSO_3H \xrightarrow[\text{hydrolysis}]{H_2O} H_3C-CH_2-OH$$

Overall: $CH_2{=}CH_2 + H_2O \xrightarrow{\text{conc. } H_2SO_4} CH_3CH_2OH$

Overall this is a *hydration* reaction, in which water is added across a multiple bond.

Propene

The acid-catalysed hydration of propene gives propan-2-ol. First, protonation gives the more stable, secondary carbocation (see p. 60) which then forms the secondary alcohol. The mechanism is similar to the addition of HBr.

$$H_3C{-}CH{=}CH_2 + H{-}OSO_3H \longrightarrow H_3C{-}\overset{+}{C}H{-}CH_3 \xrightarrow{HSO_4^-} H_3C{-}CH(OSO_3H){-}CH_3 \xrightarrow{H_2O} H_3C{-}CH(OH){-}CH_3 + H_2SO_4$$

The hydration of ethene and of propene are of great industrial importance, as the two alcohols are used on an enormous scale as solvents and as chemical feedstock.

5.4 Addition of halogens to alkenes

Ethene

Bromine and chlorine molecules are easily polarized—by glass, water, or other molecules, even by the high electron density of the C=C in the alkenes themselves.

$$Br{-}Br + Br{-}Br \longrightarrow \overset{\delta+}{Br}{-}\overset{\delta-}{Br}\cdots\overset{\delta+}{Br}{-}\overset{\delta-}{Br}$$

The electron-deficient δ+ end can add to an alkene at the same time as bromide ion is lost. The intermediate is a cyclic, three-membered ring *bromonium ion*. This bromonium ion then reacts with bromide ion to complete the overall addition of Br_2 to the double bond.

Bromide does *not* ionise spontaneously; these mechanisms are written using whole bromide molecules.

$$H_2C{=}CH_2 + Br{-}Br \longrightarrow \text{bromonium ion } (Br^+ \text{ bridging the two C}) + :\!Br\!:^- \xrightarrow{\text{substitution at carbon}} BrH_2C{-}CH_2Br$$

Overall: $H_2C{=}CH_2 + Br_2 \longrightarrow CH_2BrCH_2Br$

The formation of the bromonium ion intermediate can be understood if we look at a possible carbocation which might be formed. This carbocation can stabilize itself by sharing a nonbonded pair from the newly bonded bromine atom, to form the three-membered ring bromonium ion.

In the addition of HBr to alkenes, the added H atom has *no* non-bonded pairs of electrons and so cannot form a cyclic ion like the bromonium ion. Do you think that Cl is more (or less) likely than Br to form a cyclic 'onium ion during halogen addition to alkenes?

$$H_2C{=}CH_2 + Br{-}Br \xrightarrow{??} H_2C(Br){-}\overset{+}{C}H_2 \text{ (possible carbocation)} \quad \boxed{\text{OR}} \quad \text{bromonium ion (or)}$$

possible carbocation — OR — bromonium ion

The addition of bromine is the classical test for unsaturation. The brown colour of bromine disappears almost immediately as the alkene is converted into the colourless dibromoalkane.

If water is used as solvent for the bromine ('bromine water') then H_2O can also trap the intermediate bromonium ion to give the bromohydrin, CH_2BrCH_2OH (compound 2). If the bromine addition is carried out in a solution containing a high concentration of NaCl, then the two nucleophiles Br^- and Cl^- compete for the intermediate and some CH_2BrCH_2Cl (compound 3) will be formed as well as the dibromide (compound 1).

Carefully draw this mechanism in 3-D for the addition of Br_2 to cyclopentene, and show that the product consists of only one geometric isomer. (If you find this easy, show why two stereoisomers are produced!)

$H_2C{=}CH_2 \xrightarrow{Br_2} H_2C{-}CH_2$ (bridged Br^+)

- $\xrightarrow{Br^-}$ $BrH_2C{-}CH_2Br$ (1)
- $\xrightarrow[-H^+]{H_2O}$ $BrH_2C{-}CH_2OH$ (2)
- $\xrightarrow{Cl^-}$ $BrH_2C{-}CH_2Cl$ (3)

The formation of compounds (2) and (3) is seen as *evidence for the cationic intermediate*. What other product might be formed if sodium ethanoate ($CH_3CO_2^-Na^+$) were present in the bromine solution?

Propene

In the addition of bromine to propene, the intermediate cation is not symmetrical and more positive charge is on the secondary carbon atom than on the primary carbon atom (carbocation stability: tertiary > secondary > primary, see p. 60).

$CH_3CH{=}CH_2 + Br{-}Br \longrightarrow$ H_3C, H $\overset{\delta+}{C}{-}C$ H, H, bridged $Br^{\delta+}$ $+ Br^-$

The nucleophiles attack the more positive carbon atom ($\delta+$).

Nu^- + H_3C, H $\overset{\delta+}{C}{-}C$ H, H (bridged $Br^{\delta+}$) $\longrightarrow$ Nu, H, CH_3 C–C H, H, Br

Examples. Write the detailed mechanisms for the overall reactions.

$$Br_2 + CH_3CH{=}CH_2 \longrightarrow CH_3CHBrCH_2Br$$

$$Br_2 + CH_3CH{=}CH_2 + H_2O \longrightarrow CH_3CHOHCH_2Br + HBr$$

Chlorine and fluorine

Chlorine also adds to alkenes by an electrophilic mechanism. Fluorine reacts too violently to be useful; the initial formation of the stable C—F bond is very exothermic.

5.5 Cationic polymerization of alkenes

A Ziegler catalyst is electron-deficient and can accept a pair of electrons from the C=C bond to form a carbocation. A mixture of $TiCl_4$ and $Al(CH_2CH_3)_2Cl$ is an example of a Zeigler catalyst.

Many alkenes can be polymerized by heating with an acid catalyst, HX, or a Ziegler catalyst. The intermediate is an electron-deficient carbocation. This reacts with more of the electron-rich alkene to give another, longer carbocation, and so on, to form polymers, e.g. ethene to poly(ethene), 'polythene'.

$$H_2C{=}CH_2 + H^+ \longrightarrow H_3C{-}\overset{+}{C}H_2 + H_2C{=}CH_2 \longrightarrow H_3C{-}CH_2{-}CH_2{-}\overset{+}{C}H_2 + H_2C{=}CH_2 \longrightarrow H_3C{-}CH_2{-}CH_2{-}CH_2{-}CH_2{-}\overset{+}{C}H_2 \xrightarrow{\text{etc.}} {-}\!(CH_2{-}CH_2)_n\!{-}$$

Overall: $n\ H_2C{=}CH_2 \longrightarrow {-}\!(CH_2{-}CH_2)_n\!{-}$ poly(ethene)

If we start with propene, $CH_3CH{=}CH_2$, we keep forming secondary (rather than primary) carbocations because secondary carbocations are more stable (see p. 60).

$$H_2C{=}CH(CH_3) + H^+ \longrightarrow H_3C{-}\overset{+}{C}H(CH_3) + H_2C{=}CH(CH_3) \longrightarrow H_3C{-}CH(CH_3){-}CH_2{-}\overset{+}{C}H(CH_3) + H_2C{=}CH(CH_3) \longrightarrow H_3C{-}CH(CH_3){-}CH_2{-}CH(CH_3){-}CH_2{-}\overset{+}{C}H(CH_3) \xrightarrow{\text{etc.}} {-}\!(CH_2CHCH_3)_n\!{-}$$

Overall: $n\ H_2C{=}CHCH_3 \longrightarrow {-}\!(CH_2CHCH_3)_n\!{-}$ poly(propene)

The repeating unit of poly(propene), 'polypropylene', is $-CH_2-CHCH_3-$. In the same way phenylethene can be polymerized to poly(phenylethene), 'polystyrene', $(CH_2CHC_6H_5)_n$, *via* the cations stabilized by delocalization with the adjacent benzene ring (see p. 46).

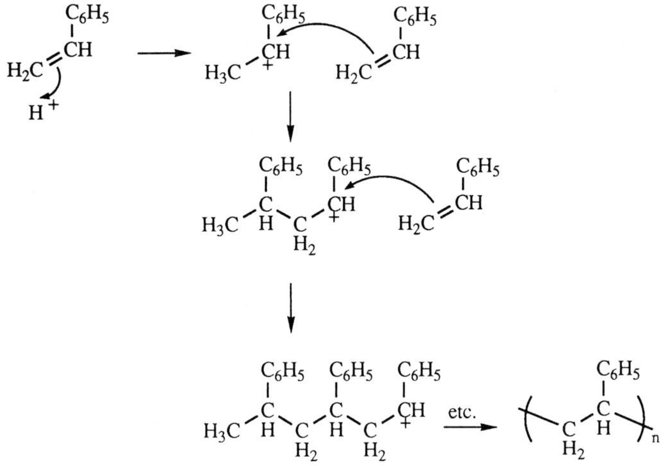

Overall: $n\ H_2C{=}CHC_6H_5 \longrightarrow -(CH_2CHC_6H_5)_n-$ poly(phenylethene)

The structures of these alkene polymers can be predicted using the stability of the cationic intermediates, in the same way as we predicted the orientation of addition of HBr to propene (see p. 60). Draw the mechanism and the product for the cationic polymerization of $H_2C{=}C(CH_3)_2$.

The polymers no longer have alkene C=C bonds, so they are resistant to attack by electrophiles like acids and $KMnO_4$. They are generally inert, like alkanes, as you would expect from their structural formulae. The reason why so many of these plastics are not biodegradable is that their molecules are not accepted as substrates by fungal or bacterial enzymes.

Rubber—a natural polyalkene.

Rubber is a natural polymer of 2-methylbuta-1,3-diene.

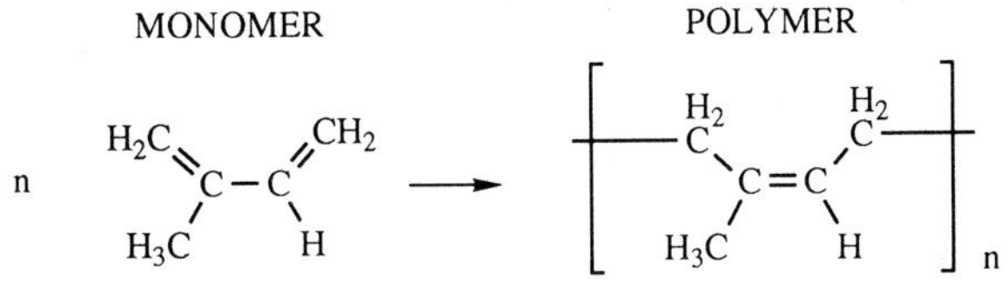

MONOMER POLYMER

Polymerization occurs in the plant to give rubber latex, a colloidal suspension of rubber in water, which is collected by 'tapping' the rubber tree. Synthetic rubber was first made from the diene monomer in 1955. Rubber still has alkene double bonds and so it can be chemically modified. Its reaction with sulphur to give a crosslinked, harder rubber is a process called *vulcanization*.

We are not giving details of these complex mechanisms (see further reading for more information).

5.6 Oxidation of alkenes

Ozonolysis

The full mechanism for ozonolysis of alkenes is complicated. The stable addition product (an ozonide) from ozone and ethene has a five-membered ring; the double bond in ethene has been broken completely. The ozonide is hydrolysed reductively (e.g. using $Zn + H_2O$) to give the aldehyde, methanal.

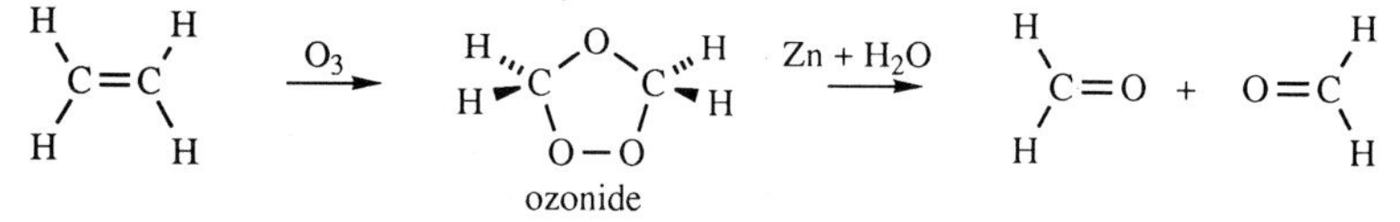

$$H_2C{=}CH_2 \xrightarrow{O_3} \text{ozonide} \xrightarrow{Zn + H_2O} H_2C{=}O + O{=}CH_2$$

ozonide

Ozonolysis of propene gives a 1 : 1 mixture of ethanal and methanal.

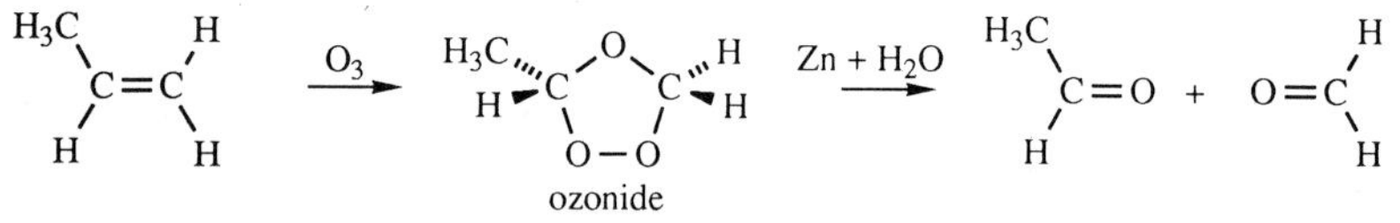

$$H_3C(H)C{=}CH_2 \xrightarrow{O_3} \text{ozonide} \xrightarrow{Zn + H_2O} H_3C(H)C{=}O + O{=}CH_2$$

ozonide

Two problems:

1. Deduce the structure of OLEIC ACID (from olive oil) which gives $CH_3(CH_2)_7CHO$ and $OHC(CH_2)_7COOH$ after ozonolysis and reductive hydrolysis.

2. Deduce the structures of the products from treating LIMONENE with first excess O_3 and then $Zn + H_2O$.

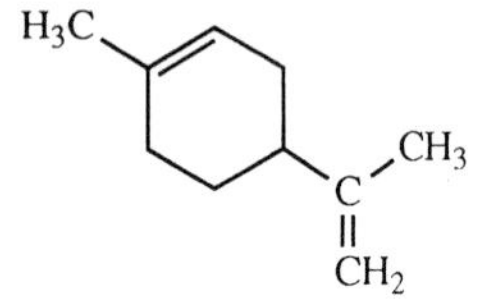

H_3C, CH_3, C, CH_2

Ozonolysis has been used to identify the substituents on each end of a C=C bond; the two carbonyl compounds produced could be identified (for example) from the melting points of their 2,4-dinitrophenylhydrazones (see p. 49). In modern organic chemistry, ozonolysis is more likely to be used to prepare aldehydes or ketones from alkenes.

Potassium manganate(VII)

Alkenes can be oxidized by potassium manganate(VII), $KMnO_4$. When an alkene is shaken with purple $KMnO_4$, the colour disappears; this can be used as a test for alkenes (like the decolorization of bromine). The alkene is oxidized to the 1,2 diol and the manganese is reduced, its oxidation state going from +7 (purple $KMnO_4$) to +4 (brown precipitate of MnO_2).

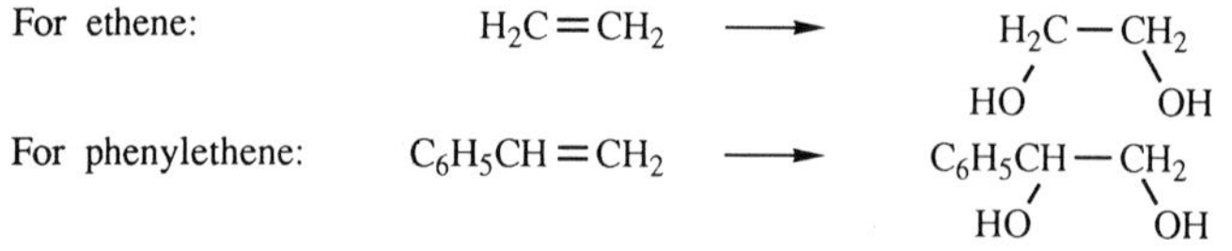

For ethene: $H_2C{=}CH_2 \longrightarrow HOH_2C{-}CH_2OH$

For phenylethene: $C_6H_5CH{=}CH_2 \longrightarrow C_6H_5CH(OH){-}CH_2OH$

What is the change in the oxidation state of Mn in going from $KMnO_4$ to the cyclic intermediate below?

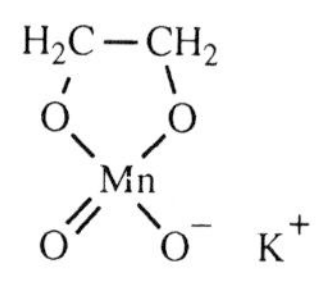

$H_2C{-}CH_2$, O, O, Mn, O, O^- K^+

Unfortunately, alcohols and diols are further oxidized by $KMnO_4$ so that this is not usually a good preparation of diols. Notice that in the phenylethene reaction only the alkene group is oxidized, not the benzene ring (see p. 74).

5.7 Benzene and related compounds

The structure of benzene

Benzene, C_6H_6, has a planar structure which is drawn in several ways.

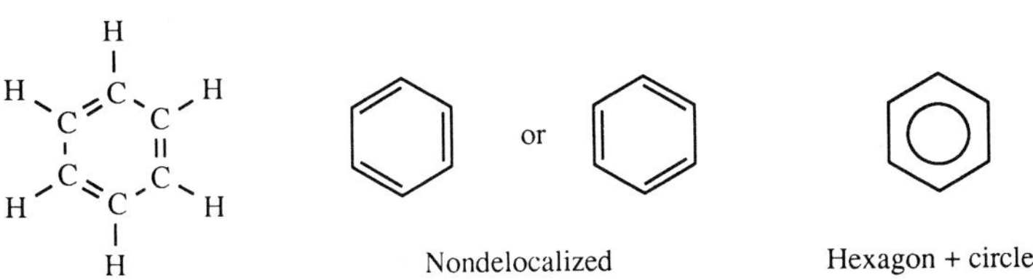

In the resonance description of its structure, benzene can be considered as a 'hybrid' of two contributing non-delocalized forms—but remember that neither of these two forms actually exists.

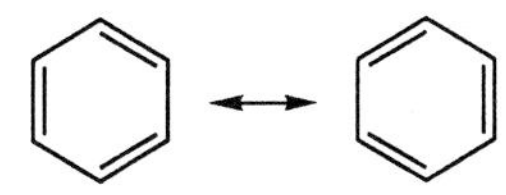

The C—H bonds are left out of the last three diagrams for clarity, as is usually done for complex organic molecules. Don't forget that if no substituent is shown on a benzene ring, you must assume that there is a hydrogen atom attached to each carbon, making C—C—H bond angles of 120°.

The actual shape of the benzene molecule is a flat, regular hexagon (from the X-ray crystal structure analysis of the solid). The carbon–carbon bond lengths are all the same (0.139 nm), a value in between normal double (0.134 nm) and single (0.154 nm) bonds.

If each carbon atom in benzene makes single bonds to each carbon neighbour and a third single bond to hydrogen, there will be one unused outer shell electron left in a p orbital on each carbon (diagram A).

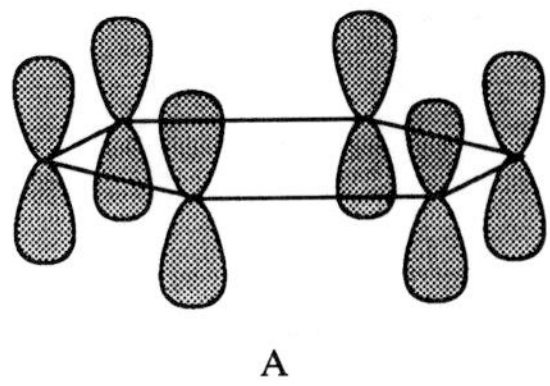

A

These six atomic orbitals can combine to form *a new set of molecular orbitals in which all six ring carbon atoms are involved*. It is this lower-energy arrangement of the remaining six electrons in their special set of molecular orbitals which gives benzene (and other 'aromatic' compounds) their characteristic properties, such as bond lengths, stability, and resistance to reaction compared with alkenes. The six electrons are delocalized over the six-carbon ring, with the electron density greatest above and below the ring in a 'double doughnut' shape (diagram B). This structure is much more stable than the theoretical 'cyclohexatriene' structure which would have alternating double and single carbon–carbon bonds. Let's look at the evidence for this extra stability, compared with alkenes.

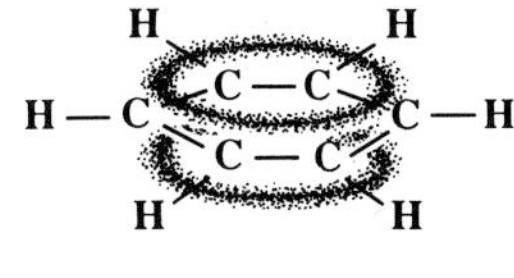

B

Remember that *six* atomic orbitals combine to make *six* molecular orbitals, of which three are bonding orbitals and are occupied (see margin on p. 3).

1. Thermochemistry. The enthalpy change on hydrogenation or on combustion of benzene is less exothermic than 'expected' in terms of a structure with three alkene C=C bonds.

(i) Hydrogenation

$$C_6H_{6(l)} + 3H_{2(g)} = C_6H_{12(l)}$$

$$\Delta H = -209.2\ kJ\ mol^{-1}$$

'expected' value is $-360.4\ kJ\ mol^{-1}$

This can be shown on a diagram.

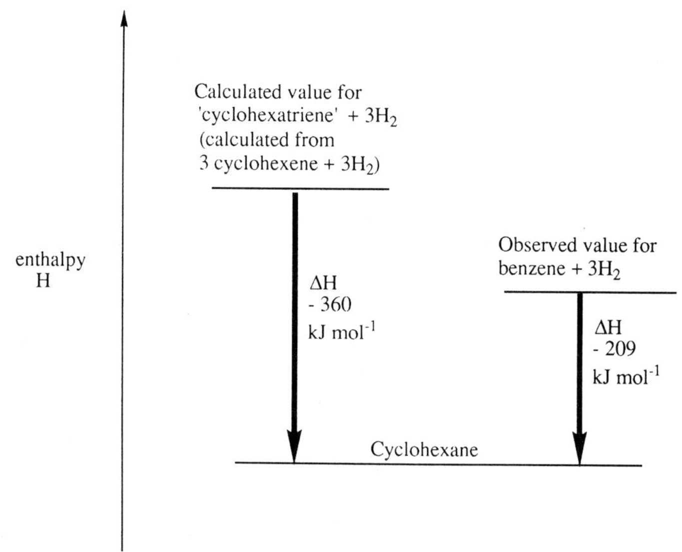

(ii) Combustion

Draw your own enthalpy diagram for the combustion.

$$C_6H_{6(l)} + 15/2\ O_{2(g)} = 6CO_{2(g)} + 3H_2O_{(l)}$$
$$\Delta H = -3313.8\ \text{kJ mol}^{-1}$$
$$\text{'expected' value is } -3473.4\ \text{kJ mol}^{-1}$$

These figures suggest that benzene is *much* more stable (of *lower* energy) than the theoretical 'cyclohexatriene' model, by about 150–160 kJ mol^{-1}. The delocalization energy contributes to this stabilization. Breaking the bonds in benzene is more *endo*thermic than it would be for cyclohexatriene; the difference is used as a measure of the delocalization energy.

2. *Low reactivity of benzene.* If benzene is *more stable* than the alkenes, we expect it to be *less reactive* (more energy needed for reaction). This is true for a wide range of reactions. Here are some examples.

(i) *Hydrogenation.* Cyclohexene can be hydrogenated by hydrogen and a Ni catalyst at room temperature and pressure; benzene requires 200°C and 200 kP (see p. 78).

(ii) *Bromination.* No catalyst is need for the reaction of bromine with alkenes, but bromination of benzene needs a metal bromide (see p. 71). Both begin by an electrophilic addition reaction.

(iii) *Polymerization.* Phenylethene has both a benzene ring and an alkene C=C, but when it is polymerized with acid or radicals only the alkene C=C reacts (see pp. 65 and 81).

How do we draw benzene?

The two standard ways are the 'nondelocalized' and 'hexagon + circle' diagrams:

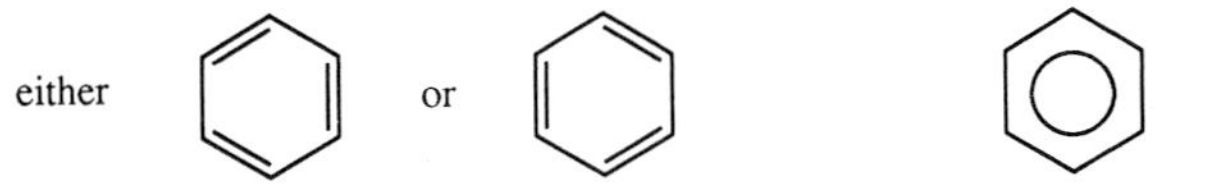

either or

Nondelocalized Hexagon + circle

First of all, neither way is 'right' but both are useful. A 'nondelocalized' diagram implies separate double and single carbon–carbon bonds; this is not so, as all the carbon–carbon bond lengths are identical in benzene. Nevertheless, these diagrams are useful for drawing mechanisms, and we will use them for mechanisms. In the hexagon + circle diagram, the circle represents the six delocalized electrons in their set of molecular orbitals. This shows the hexagonal symmetry of benzene but it is more difficult to use when drawing mechanisms. Professional chemists often use the hexagon + circle for speed but change to nondelocalized diagrams when drawing mechanisms.

Only *one* 1,2 disubstituted $C_6H_4X_2$ exists; that is

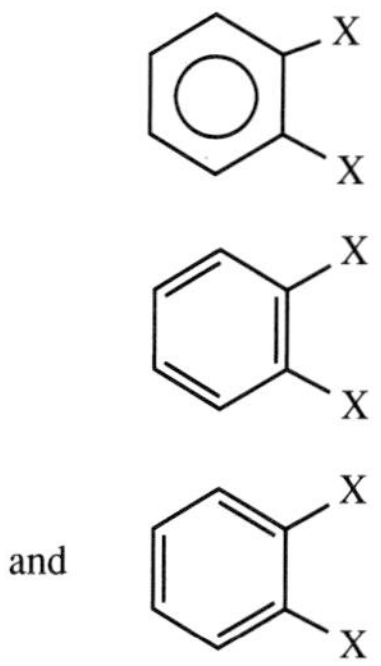

X X X X and X X

all represent the same compound.

5.8 Electrophilic substitution of benzene

Benzene reacts with electrophiles through the high electron density associated with the set of six delocalized electrons. This is similar to the way in which alkenes react through their π-bond electrons. Both react first by addition of the electrophile, but instead of completing the addition as the alkenes do, the benzene intermediate loses a proton, H^+, to give overall substitution. In terms of energy, it is more worthwhile to regain the special stability from delocalization than to make the extra bond and leave a diene. Using $^{\delta+}E{-}X^{\delta-}$ as a general electrophile:

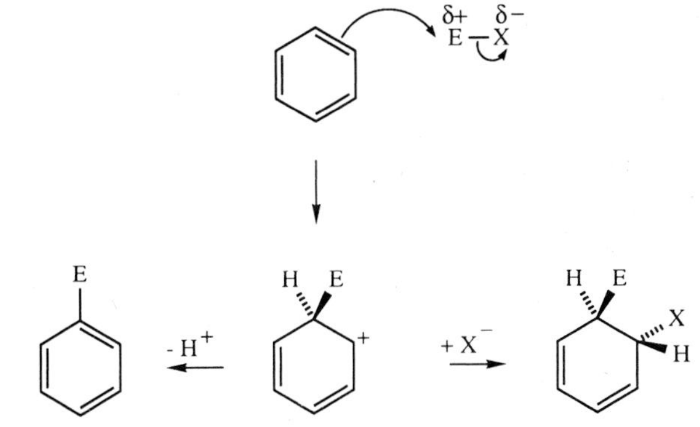

δ+ δ− E X

E H E H E X H

$-H^+$ $+X^-$

substitution product

addition product

The stability of the benzene system pulls these reactions towards overall substitution rather than addition. The three electrophilic substitution reactions we will look at in more detail are nitration (in which NO_2 replaces H), bromination (Br replaces H), and chlorination (Cl replaces H).

Nitration

When benzene is warmed with a mixture of concentrated nitric acid and concentrated sulphuric acid, nitrobenzene is formed.

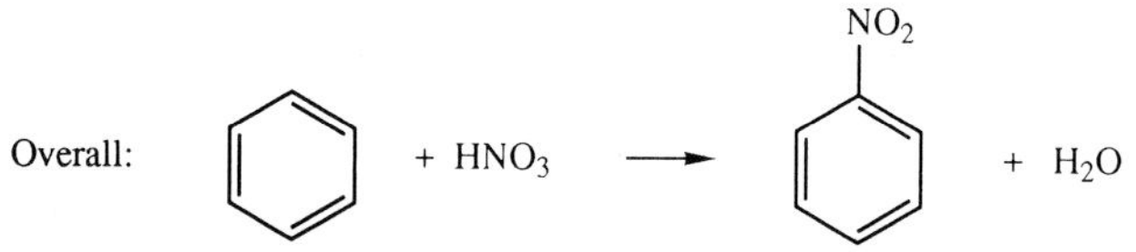

Overall: benzene + HNO_3 ⟶ nitrobenzene + H_2O

The detailed equilibria are:

$$H_2SO_4 + HNO_3 \rightleftharpoons HSO_4^- + H_2NO_3^+$$

$$H_2NO_3^+ \rightleftharpoons H_2O + NO_2^+$$

$$H_2O + H_2SO_4 \rightleftharpoons H_3O^+ + HSO_4^-$$

The H_2SO_4 does not appear in the equation. It acts as an acid catalyst and the reaction is very slow without it. The electrophile is the electron-deficient ion NO_2^+ ($O{=}\overset{+}{N}{=}O$). This is formed by mixing a very strong acid (e.g. H_2SO_4) with nitric acid.

$$\text{Overall:}\quad 2H_2SO_4 + HNO_3 \longrightarrow \overset{+}{N}O_2 + H_3O^+ + 2HSO_4^-$$

We will now draw the mechanism of the substitution using a non-delocalized diagram for benzene. This makes it easy to compare this mechanism with the electrophilic addition to alkenes.

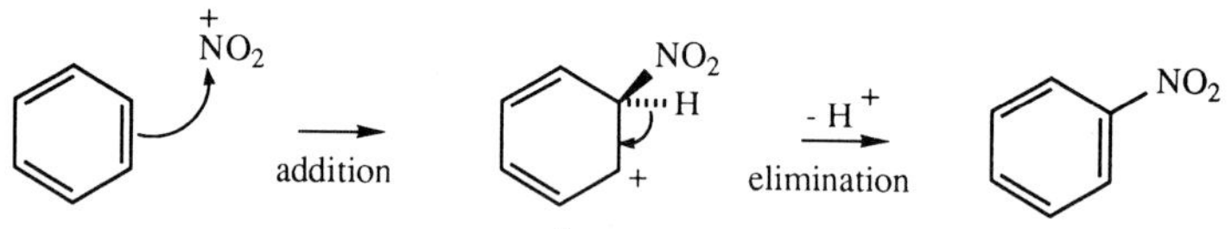

$\overset{+}{N}O_2$ — addition → A (NO_2, H, +) — $-H^+$ elimination → nitrobenzene (NO_2)

Compare:

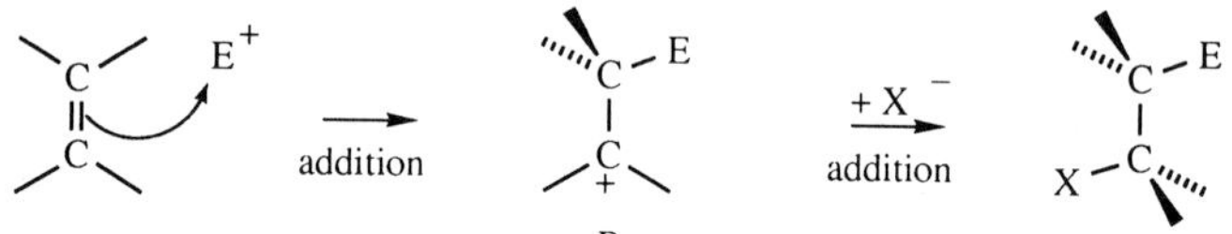

E^+ — addition → B (E, C, C+) — $+X^-$ addition → (E, C, C, X)

Cation A:

Intermediate cations A and B are formed by the initial addition of E^+, but notice that the positive charge in cation A can be delocalized using the two remaining conjugated double bonds. This can be drawn as a 'horseshoe' to show the delocalized electrons. Note that the 'horseshoe' delocalization involves carbons 2 to 6 inclusive *but not carbon 1* (which is saturated and tetrahedral). The reaction is completed by loss of a proton to re-form the fully delocalized benzene system.

$\overset{+}{N}O_2$ ⟶ (NO_2, H, +) — $-H^+$ → (NO_2)

or

$\overset{+}{N}O_2$ ⟶ (NO_2, H, +) — $-H^+$ → (NO_2)

$$\text{Overall:}\quad C_6H_6 + HNO_3 \longrightarrow C_6H_5NO_2 + H_2O$$

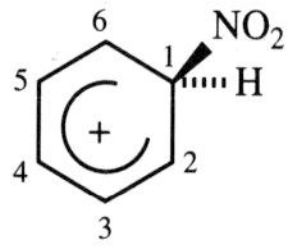

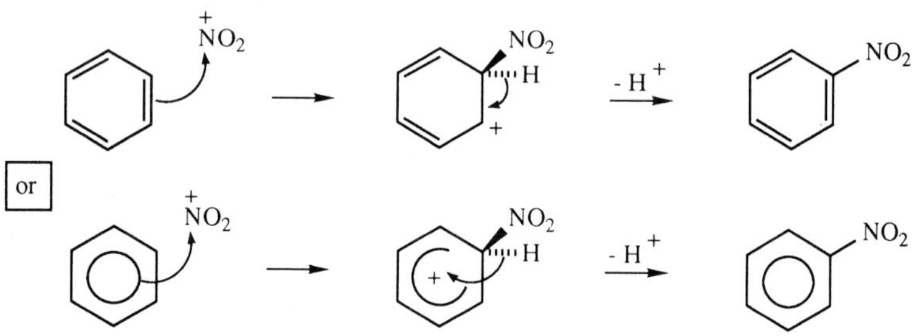

The structure of the nitro-group can be drawn as

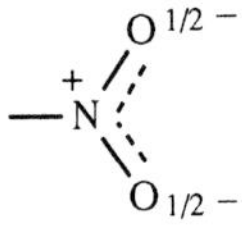

$O^{1/2-}$, $-\overset{+}{N}$, $O^{1/2-}$

which is also delocalized.

Use of nitration reactions. Nitration is an important substitution reaction because nitrocompounds are easily reduced to amines.

$$C_6H_5NO_2 + 3H_2 \xrightarrow{\text{Pd catalyst}} C_6H_5NH_2 + 2H_2O$$

The aromatic amines are very useful synthetic intermediates because a wide variety of new functional groups can be introduced *via* diazotization (see p. 73).

The explosive TNT (trinitrotoluene, or 2,4,6-trinitromethylbenzene) is made by (careful!) nitration of methylbenzene.

$$C_6H_5CH_3 + 3HNO_3 \xrightarrow{\text{conc. } H_2SO_4} CH_3C_6H_2(NO_2)_3 + 3H_2O$$

Polynitrocompounds are thermodynamically unstable chiefly because they produce the very stable nitrogen molecule, N_2, when they decompose.

PARATHION is an aromatic nitro-compound which is a broad spectrum insecticide. It mimics acetylcholine and acts as an inactivator of the enzyme acetylcholine esterase.

Bromination

Ethene decolorizes a brown bromine solution rapidly to form colourless 1,2-dibromethane by an addition reaction.

$$H_2C{=}CH_2 + Br_2 \longrightarrow BrCH_2CH_2Br$$

If you shake benzene with bromine solution, nothing happens. Bromine needs to be much more highly polarized before it can react with the more stable aromatic compound. Bromine can be polarized or even ionized using metal halides that are electron-deficient; e.g. $FeBr_3$, $AlBr_3$, $AlCl_3$. These can accept a bromide ion from bromine to form a new ion pair containing the very reactive electrophile, Br^+.

Why are $FeBr_3$ and $AlCl_3$ described as 'electron-deficient'? Would you expect either BBr_3 or NBr_3 to act as a catalyst for this reaction?

$$Br_2 + FeBr_3 \rightleftharpoons Br^+ + FeBr_4^-$$

$$Br_2 + AlBr_3 \rightleftharpoons Br^+ + AlBr_4^-$$

Br^+ has only *six* electrons in its outer shell and will now react with the benzene.

$$Br{:}Br + AlBr_3 \longrightarrow [Br]^+ [AlBr_4]^-$$

$$C_6H_6 + Br^+ \longrightarrow \text{C} \xrightarrow{-H^+} C_6H_5Br$$

C

Here again, the positive charge of intermediate cation C can be delocalized. When all the equations are written out, we find that the metal halide is regenerated at the end; formally, it is a *catalyst* to polarize the Br_2.

$$Br_2 + AlBr_3 \rightleftharpoons Br^+ + AlBr_4^-$$

$$C_6H_6 + Br^+ \longrightarrow C_6H_6Br^+$$

$$C_6H_6Br^+ \longrightarrow C_6H_5Br + H^+$$

$$H^+ + AlBr_4^- \rightleftharpoons HBr + AlBr_3$$

Overall: $$C_6H_6 + Br_2 \longrightarrow C_6H_5Br + HBr$$

All these reactions are usually carried out away from bright light to avoid radical reactions (see p. 78).

Chlorination

This follows a similar pathway to bromination and again needs a catalyst. Usually $FeCl_3$ or $AlCl_3$ are used with Cl_2.

Reactions of phenol and phenylamine

Halogenation of phenol, C_6H_5OH, and phenylamine, $C_6H_5NH_2$, is interesting because these two aromatic compounds do not need a catalyst and both form tri-substituted products. This suggests that phenol and phenylamine are *more* reactive to electrophiles than benzene (which *does* need a catalyst and forms a *mono*substituted compound). Nevertheless, they all still follow a similar mechanism of addition–elimination rather than addition.

The positions at which the bromines substitute in phenylamine are related to the higher electron densities caused by delocalization of the N nonbonded pair.

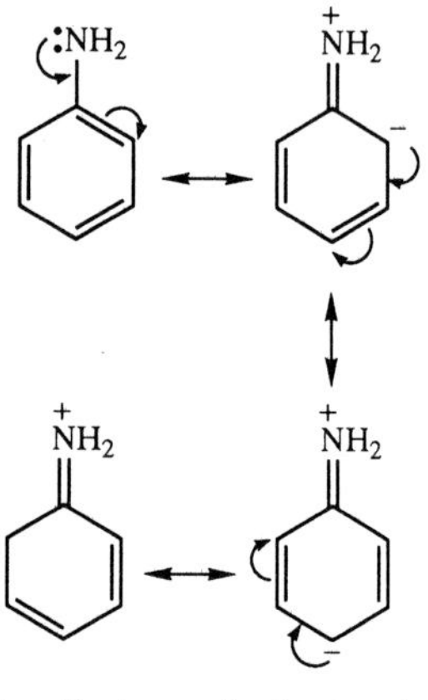

NH_2 (phenylamine) $+ 3Br_2 \longrightarrow$ 2,4,6-tribromophenylamine $+ 3HBr$

white precipitate

A similar effect operates in phenol.

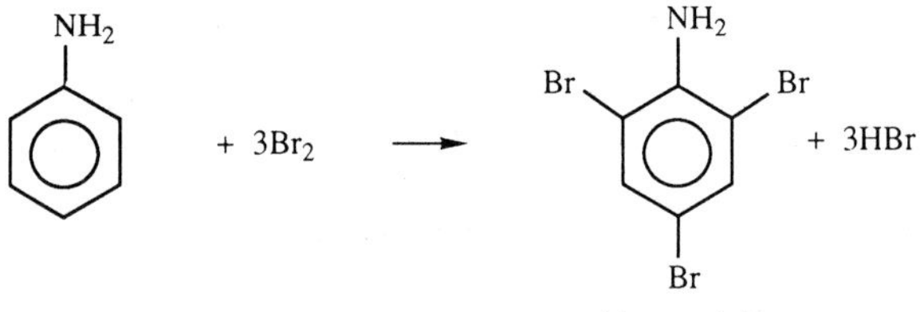

Why are phenol and phenylamine more reactive towards electrophiles than benzene? Since they are more reactive, we might expect them to have higher electron density. In phenol and phenylamine, the nonbonded pairs of electrons on the O or N atoms can be involved in the delocalized system and so raise the electron density in the benzene ring. This delocalization of the nonbonded pair of electrons in phenoxide ion is responsible for the increased acidity of phenols over alcohols (see p. 37). See also the related effect on the basicity of phenylamine (p. 38).

Phenol reacts with chlorine to give the familiar antiseptic, TCP (2,4,6-trichlorophenol).

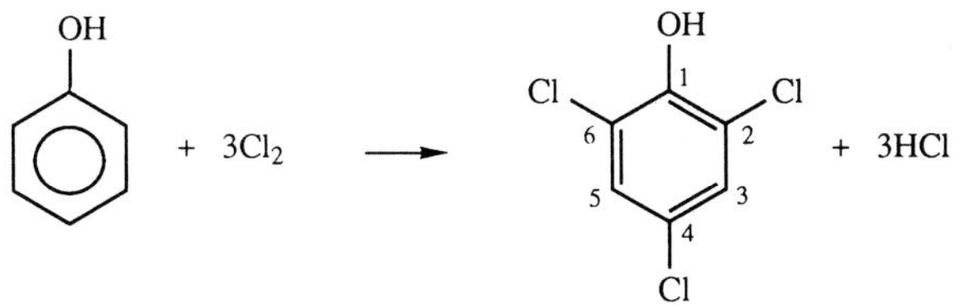

OH + $3Cl_2$ → (OH at 1; Cl at 2, 4, 6) + 3HCl

DIPRIVAN is a phenol which is a fast-acting, injectable anaesthetic for surgery.

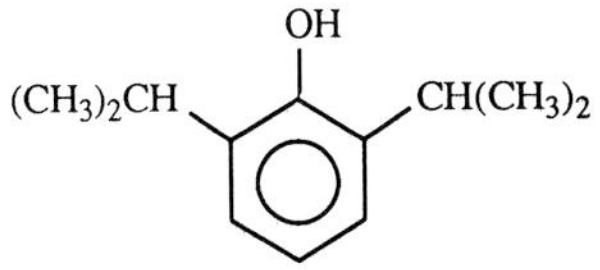

OH
$(CH_3)_2CH$ $CH(CH_3)_2$

Diazotization and diazo-coupling

One of the most important functional groups in the chemistry of aromatic compounds is the amino group. This is because it can be converted into the $-N_2^+$ group, a process called *diazotization.*

$$NaNO_2 + HCl \longrightarrow HNO_2 + NaCl$$

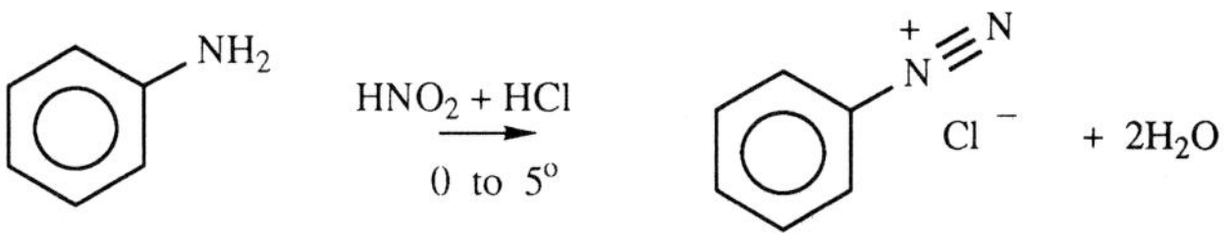

NH_2 $\xrightarrow[0\text{ to }5^o]{HNO_2 + HCl}$ $\overset{+}{N}\equiv N$ Cl^- + $2H_2O$

a diazonium salt

The $-N_2^+$ group is weakly electrophilic and can react with very electron-rich aromatic compounds such as phenols. This is the basis of the azo dye industry, because the 'aromatic ring-N=N-aromatic ring' products are brightly coloured, fast azo dyes. They are widely used in the printing industry.

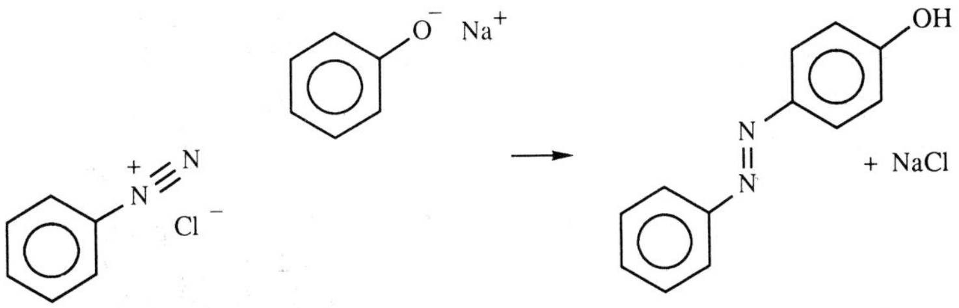

$\overset{-}{O}$ Na^+

$\overset{+}{N}\equiv N$ Cl^- → N=N, OH + NaCl

Draw the two geometric isomers of the dye.

Diazonium salt + sodium phenoxide → an azo dye

Work out the mechanism of this last reaction (it's an electrophilic substitution reaction on the phenoxide ion).

The $-N_2^+$ of the diazonium salt also provides the best leaving group of all, stable molecular nitrogen, N_2. Diazonium salts are also useful compounds for further synthesis because many different groups can be substituted for the N_2^+, e.g. cyanide.

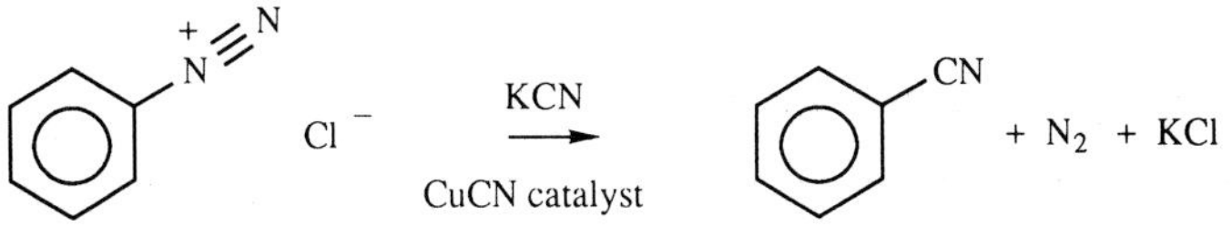

$\overset{+}{N}\equiv N$ Cl^- $\xrightarrow[\text{CuCN catalyst}]{KCN}$ CN + N_2 + KCl

Conclusions: general mechanism

The general mechanism for the reaction of benzene with an electrophile E^+ is

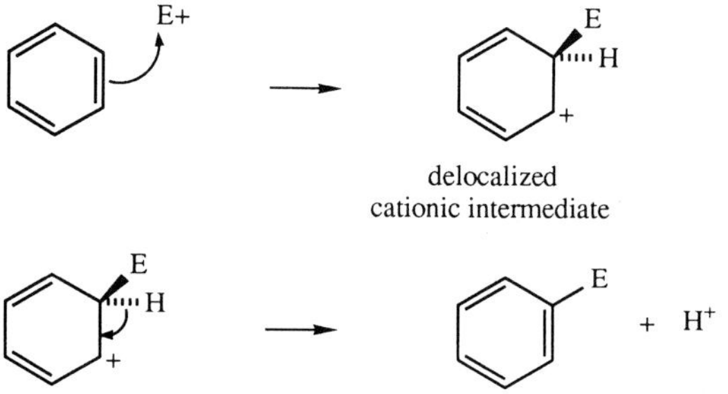

E+

E

H

+

delocalized
cationic intermediate

E

H

+

E

+ H^+

The stable benzene molecule needs a reactive (unstable) electrophile, which is often made in the reaction mixture, e.g. NO_2^+.

$$HNO_3 + 2H_2SO_4 \longrightarrow NO_2^+ + H_3O^+ + 2HSO_4^-$$

5.9 Oxidation of benzene and related compounds with $KMnO_4$

Benzene is inert to $KMnO_4$. In special conditions it can even be used as a solvent for $KMnO_4$. Contrast this with the rapid oxidation of alkenes (see p. 66). Side-chains on benzene rings may be oxidized. Methylbenzene is oxidized by $KMnO_4$ to benzoic acid, but the benzene ring remains intact.

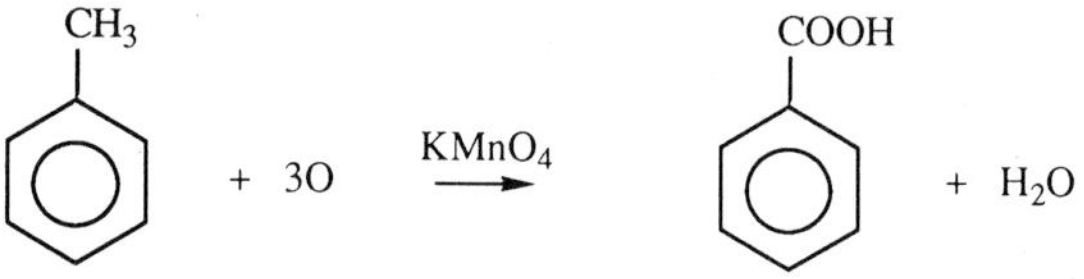

CH_3

+ 3O $\xrightarrow{KMnO_4}$

COOH

+ H_2O

5.10 Benzene compared with alkenes

Some points to note are:

1. Both are electron-rich and react with electrophiles.
2. The bonding in benzene is different from alkenes; evidence: C—C bond lengths (see p. 67).
3. Benzene and related compounds are thermodynamically much more stable than expected for a triene; evidence: heats of hydrogenation and combustion (see p. 67).
4. Alkenes undergo *addition* reactions whereas benzene undergoes overall *substitution*; evidence: see bromination (pp. 62 and 71).
5. Benzene and related compounds are more resistant to oxidation and reduction; see $KMnO_4$ (p. 66) and catalytic hydrogenation (p. 78).

6 Reactions with radical intermediates

6.1 Introduction

In Chapters 4 and 5 the reaction mechanisms all involved the transfer of pairs of electrons. These are heterolytic reactions and often have charged intermediates. In this chapter we will be looking at *homolytic reactions* which involve *radical* intermediates. A radical has an odd *unpaired electron*, and is normally uncharged. Radicals are electron-deficient in the sense that the atom carrying the unpaired electron is one electron short of a stable outer shell, such as carbon or chlorine with seven electrons instead of eight. For drawing mechanisms involving single electron transfers we use *fishhook arrows* (see p. 20).

Formation and reactions of radicals

In order to unpair two bonding electrons to make two radicals we need to supply energy, usually as heat or light (often high-energy ultraviolet light). The formation of radicals at the beginning of a reaction is called the *initiation*. The reaction then continues by the *propagation* steps, in which a radical reacts with a molecule to give another radical as one of the products. The propagation steps can go on and on to make a *chain reaction*, until two radicals combine to form a stable compound; this is the *termination* reaction. We will now look at these three processes, *initiation*, *propagation*, and *termination*, for the radical chlorination of methane.

6.2 Halogenation of hydrocarbons

The radical chlorination of methane

This is a chain reaction.

Initiation. Chlorine and methane are mixed and either irradiated with ultraviolet light or left in sunlight, of which ultraviolet light forms a part. The energy of the light is absorbed by the chlorine molecules, which split *homolytically* to give two radicals. It is a photochemical reaction.

$$\mathrm{Cl{-}Cl} \xrightarrow{\text{light}} \mathrm{Cl^{\bullet}} + \mathrm{{}^{\bullet}Cl}$$

This is energetically easier than splitting the stronger $H{-}CH_3$ bond of methane.

Propagation. Each chlorine radical takes a hydrogen atom from methane to produce HCl and a new radical $^{\bullet}CH_3$, in which the C atom has a share in only 7 outer shell electrons.

$$CH_4 + \dot{Cl} \longrightarrow {}^{\bullet}CH_3 + HCl$$

or

$$H_3C{-}H \;\; {}^{\bullet}Cl \longrightarrow H_3C^{\bullet} + H{-}Cl$$

or

$$H:\underset{H}{\overset{H}{C}}:H \;\; {}^{\bullet}\ddot{Cl}: \longrightarrow H:\underset{H}{\overset{H}{C}}{}^{\bullet} + H:\ddot{Cl}:$$

The new methyl radical $^{\bullet}CH_3$ can then take a chlorine atom from Cl_2 to make CH_3Cl and another chlorine radical.

$$Cl_2 + {}^{\bullet}CH_3 \longrightarrow {}^{\bullet}Cl + CH_3Cl$$

or

$$Cl{-}Cl \;\; {}^{\bullet}CH_3 \longrightarrow Cl^{\bullet} + Cl{-}CH_3$$

This chlorine radical can go on to react with another molecule of methane, and so on. Notice that in each of these propagation steps there is an odd number of electrons on both sides of the equation.

Termination. The propagation chain is broken when any of the radicals involved reacts to form only electron-paired compounds. Here the propagation radicals are $Cl^{\bullet}$ and $^{\bullet}CH_3$, so that the most likely chain termination reactions are combinations of these radicals.

$$Cl^{\bullet} \;\; {}^{\bullet}Cl \longrightarrow Cl{-}Cl \quad \text{(the reverse of initiation)}$$

$$Cl^{\bullet} \;\; {}^{\bullet}CH_3 \longrightarrow Cl{-}CH_3$$

$$H_3C^{\bullet} \;\; {}^{\bullet}CH_3 \longrightarrow H_3C{-}CH_3$$

Putting all the steps together:

Initation:	Cl_2	$\xrightarrow{\text{light}}$	$2Cl^{\bullet}$
Propagation:	$CH_4 + Cl^{\bullet}$	$\longrightarrow$	$^{\bullet}CH_3 + HCl$
	$Cl_2 + {}^{\bullet}CH_3$	$\longrightarrow$	$Cl^{\bullet} + Cl\,CH_3$
Termination: e.g.	$2\,CH_3^{\bullet}$	$\longrightarrow$	H_3CCH_3
OVERALL:	$Cl_2 + CH_4$	$\xrightarrow{\text{light}}$	$CH_3Cl + HCl$

One of the key stages in the synthesis of the herbicide PARAQUAT

$$H_3C{-}\overset{+}{N}\text{(pyridinium)}{-}\text{(pyridinium)}\overset{+}{N}{-}CH_3 \quad Cl^- \quad Cl^-$$

is the combination of two radicals. Paraquat is used for cleaning cereal stubble before direct drilling, as part of 'minimum cultivation' to retain soil structure. Paraquat is inactivated by the soil.

We can also show this as a radical flow diagram.

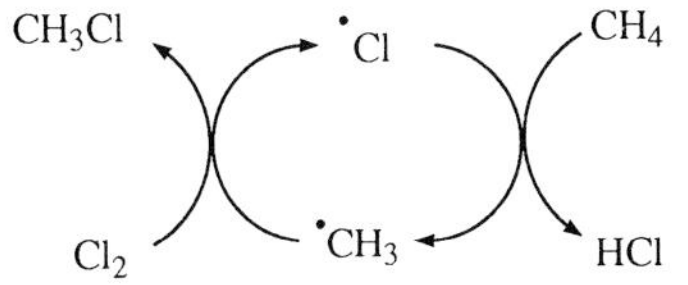

The central circle contains the chain-carrying radicals in the propagation steps. The side arrows show which starting materials go *in* and which products go *out* at each propagation step.

$$CH_3Cl \leftarrow \cdot Cl \leftarrow CH_4$$
$$Cl_2 \rightarrow \cdot CH_3 \rightarrow HCl$$

This is the major overall reaction if there are equimolar amounts of Cl_2 and CH_4. If there is more Cl_2, the reaction can go on to give CH_2Cl_2, $CHCl_3$, and eventually CCl_4 as well as HCl. The *propagation* steps for CH_2Cl_2 are:

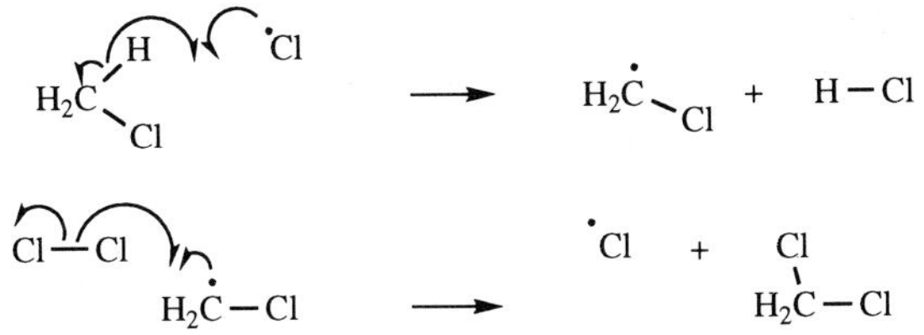

$$H_2C(H)Cl + \cdot Cl \longrightarrow H_2\dot{C}-Cl + H-Cl$$

$$Cl-Cl + H_2\dot{C}-Cl \longrightarrow \cdot Cl + H_2C(Cl)-Cl$$

Write the corresponding steps for $CHCl_3$ and CCl_4. Remember that the termination steps are any combination of chain-carrying radicals. By adjusting the relative amounts of CH_4 and Cl_2, you can make mostly CH_3Cl, CH_2Cl_2, $CHCl_3$, or CCl_4. Draw radical flow diagrams for these, too.

Bromination of methane

The photochemical bromination of methane is a very similar radical reaction. It starts with the homolytic splitting of the Br—Br bond to give two bromine radicals. The propagation steps, involving $Br^\cdot$ and $^\cdot CH_3$, follow and the termination is a combination of two radicals. Draw the mechanisms for the *initiation*, *propagation*, and *termination* steps for this reaction, using fishhook arrows. Draw a radical flow diagram for this reaction as well.

Halogenation of other hydrocarbons

Chlorination and bromination of methane are relatively simple because methane has only one type of C—H bond. Propane, $CH_3CH_2CH_3$, has methyl C—H's (six) and CH_2 C—H's (two). Two different hydrocarbon radicals can be formed, primary $CH_3CH_2CH_2^\cdot$ and secondary $CH_3\,^\cdot CHCH_3$. The secondary $CH_3\,^\cdot CHCH_3$ radical is more easily formed because the CH_2 C—H bonds are weaker than methyl C—H bonds; *but* there are three times as many methyl C—H bonds. These factors are closely balanced and mixtures of products are often obtained.

On an industrial scale it is sometimes worthwhile to separate these mixtures.

In the photochemical chlorination of methylbenzene, $C_6H_5CH_3$, the side-chain radical $C_6H_5CH_2^\cdot$ is much more stable and is more easily formed than any of the ring radicals, such as the one shown in the margin. This is because

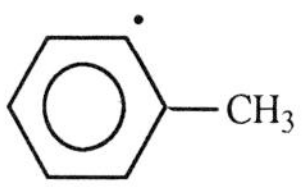

$-CH_3$

$C_6H_5CH_2^{\bullet}$ is stabilized by delocalization (we shall see this type of radical again in the polymerization of phenylethene; see p. 81).

$$\text{Overall: } C_6H_5CH_3 + Cl_2 \xrightarrow{\text{light}} C_6H_5CH_2Cl + HCl$$

Conclusions

These radical reactions are very important because an inert hydrocarbon is turned into a more reactive halocompound, allowing access to a wide range of organic reactions.

The conditions for these homolytic reactions are quite distinct from heterolytic reactions. For example, we can brominate methylbenzene to give two different products by choosing *different reaction conditions*.

Homolytic substitution:

$$C_6H_5CH_3 + Br_2 \xrightarrow{\text{light}} C_6H_5CH_2Br + HBr$$

Heterolytic substitution:

$$C_6H_5CH_3 + Br_2 \xrightarrow{FeBr_3} 4\text{-}BrC_6H_4CH_3 + HBr$$

Part of the synthesis of DICHLOBENIL (see p. 50) involves the radical chlorination of a side-chain methyl group.

$$2,6\text{-}Cl_2C_6H_3CH_3 + 2Cl_2 \rightarrow 2,6\text{-}Cl_2C_6H_3CHCl_2 + 2HCl$$

NOLVADEX is used to treat breast cancer.
Using structural formulae, draw the equation for the reaction of Nolvadex with hydrogen and a palladium catalyst at room temperature.

$$CH_3CH_2(C_6H_5)C{=}C(C_6H_5)(C_6H_4{-}O{-}CH_2CH_2{-}N(CH_3)_2)$$

6.3 Reduction: catalytic hydrogenation

Alkenes can be reduced to alkanes by the addition of hydrogen in the presence of a finely-divided transition metal catalyst, often Ni, Pd, or Pt.

$$CH_3CH{=}CH_2 + H_2 \xrightarrow[\text{room temp.}]{Pd} CH_3CH_2CH_3$$

$$C_6H_5CH{=}CH_2 + H_2 \xrightarrow[\text{room temp.}]{Pd} C_6H_5CH_2CH_3$$

The mechanism of these reactions is not fully understood. Both alkene and hydrogen are probably absorbed on the catalyst surface, and the H_2 molecule is probably split into hydrogen atoms. The reaction is thus *heterogeneous* (taking place in *more than one phase*, here gas + solid) and *homolytic* (H—H goes to $2H^{\bullet}$, bonding electrons shared evenly).

This reaction is very important industrially, for example in the conversion ('hardening') of oils into fats, such as the hydrogenation of palm oil for margarine manufacture. Nickel catalysts, being relatively cheap and robust, are commonly used. In research laboratories, where smaller amounts of more delicate chemicals are used, the more expensive and more reactive Pd or Pt catalysts are usually used.

Benzene rings are much more resistant to hydrogenation than alkenes (see the phenylethene reaction) and usually need high pressure and high temperature, even with more reactive catalysts. This is an important difference (see p. 68).

Notice that the *nucleophilic* reducing agent $LiAlH_4$ does *not* react with benzene or ethene.

6.4 The cracking of hydrocarbons

The world's major sources of hydrocarbons are coal and oil. Both contain a wide range of hydrocarbons. Some of the most desirable hydrocarbons have quite low relative molecular masses. The less useful, higher relative molecular mass hydrocarbons can be 'cracked'—that is, converted by heating into hydrocarbons of lower relative molecular mass. The heat energy causes homolytic breakage of the C—C bonds to give radicals, which react further to give alkanes and alkenes.

Plant oils are possible renewable fuels for the future. The methyl ester of rape-seed oil has been tested as an alternative tractor fuel, and palm oil has been suggested as a future car fuel.

e.g. $$CH_3(CH_2)_4CH_2-CH_2(CH_2)_4CH_3 \xrightarrow{\text{heat}} 2\ CH_3(CH_2)_4\dot{C}H_2$$

$$CH_3(CH_2)_4\dot{C}H_2 \quad H-CH(\dot{C}H_2)(CH_2)_3CH_3 \longrightarrow CH_3(CH_2)_4CH_3 + CH_2{=}CH(CH_2)_3CH_3$$

Overall: $$CH_3(CH_2)_{10}CH_3 \xrightarrow{\text{heat}} CH_3(CH_2)_4CH_3 + CH_3(CH_2)_3CH{=}CH_2$$

The high temperatures are needed because of the high C—C bond energy. Ethene, propene, and ethyne are all manufactured by the cracking of higher hydrocarbons.

Hydrogen is a by-product of some cracking reactions, e.g.

$$C_{12}H_{26} \longrightarrow C_{12}H_{24} + H_2$$

6.5 Radical polymerization of alkenes

Ethene

Radicals will add to carbon-carbon double bonds to make new radicals. If $R^{\bullet}$ represents a radical from the initiation step:

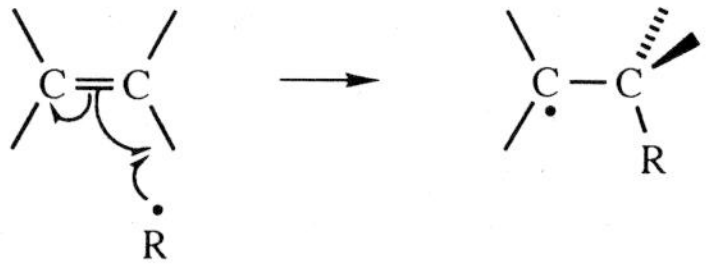

The new radical can add to another alkene molecule, and so on to make a polymer. For ethene these steps are:

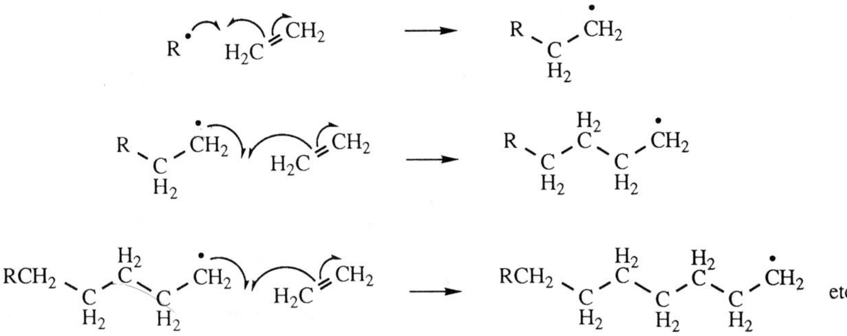

$$R^{\bullet} + H_2C{=}CH_2 \longrightarrow RCH_2\dot{C}H_2$$

$$RCH_2\dot{C}H_2 + H_2C{=}CH_2 \longrightarrow RCH_2CH_2CH_2\dot{C}H_2$$

$$RCH_2CH_2CH_2\dot{C}H_2 + H_2C{=}CH_2 \longrightarrow RCH_2CH_2CH_2CH_2CH_2\dot{C}H_2 \quad \text{etc.}$$

Overall: $nH_2C{=}CH_2 \xrightarrow{R^{\bullet}} -\!\!(CH_2CH_2)_n\!\!-$

polyethene, or polythene

Propene

Propene is not a symmetrical alkene and $R^{\bullet}$ could add to either end of the double bond to give two different radicals, one primary (P) and one secondary (S):

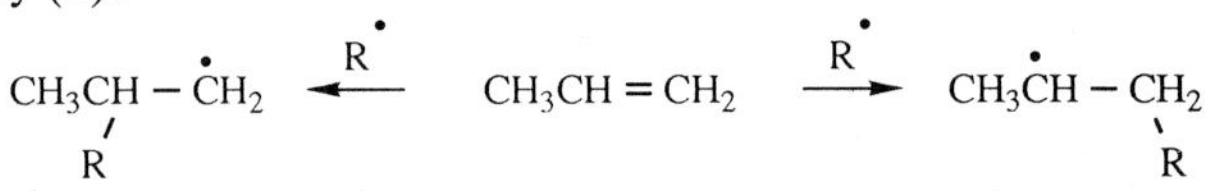

$$\underset{(P)}{CH_3CH(R)-\dot{C}H_2} \xleftarrow{R^{\bullet}} CH_3CH=CH_2 \xrightarrow{R^{\bullet}} \underset{(S)}{CH_3\dot{C}H-CH_2R}$$

Usually reaction goes *via* the more stable radical, which is preceded by the lower-energy transition state. The stabilities of alkyl radicals are in the order

tertiary > secondary > primary

e.g. $(CH_3)_3\dot{C} > (CH_3)_2\dot{C}H > CH_3\dot{C}H_2$

This is the same order as for carbocations. Both species are 3-coordinate and electron-deficient, so some of the same influences may be operating here, e.g. stereochemical and inductive effects (see p. 60). Polymerization of propene goes through secondary radicals like (S) and so gives a regular head-to-tail polymer. Here is the overall equation: you should write the detailed mechanism.

Poly(propene) is used to make ropes and in light engineering.

$$n\ CH_3CH{=}CH_2 \xrightarrow{R^{\bullet}} -\!\!(CH(CH_3)-CH_2)_n\!\!-$$

polypropene, or polypropylene

Phenylethene

When a radical adds to phenylethene, the more stable radical intermediate is the one where the carbon with the odd electron also carries the benzene ring. This radical can be stabilized by delocalization over the ring.

HC=CH₂ ·R (C₆H₅) → ·HC–CH₂–R (C₆H₅)

HC=CH₂ + ·HC–CH₂–R → ·HC–CH₂–HC–CH₂–R etc.

Overall: $nC_6H_5CH{=}CH_2 \longrightarrow -[HC(C_6H_5)-CH_2]_n-$

The polymer produced is 'polystyrene'.

Expanded polystyrene, used for thermal insulation and packaging, is a honeycomb made using gaseous pentane during the polymerization.

Initiators for the manufacture of polyalkenes

All these polymerizations need an initiating radical R˙. This is often made by heating a compound with a weak bond which will break homolytically; for example, a peroxide.

$$(CH_3)_3C-O-O-C(CH_3)_3 \xrightarrow{\text{heat}} 2\ (CH_3)_3C-\dot{O}$$

weak bond broken — INITIATOR

By careful control of the conditions of the reaction—time, temperature, pressure, concentration of monomer, nature and concentration of initiator—polymers of predictable average relative molecular mass and physical properties can be manufactured. The chemical inertness of these polymers is the reason why they are so widely used, although this means that their disposal is also a problem (see p. 65).

6.6 Enzyme-catalysed radical reactions

Higher animals and plants cannot make their own vitamin B_{12}; it is only synthesized by micro-organisms such as anaerobic bacteria. Pernicious anaemia is caused by impaired absorption of vitamin B_{12} from the ileum, and it can be cured by relatively large injections of the vitamin.

Most enzymic reactions are heterolytic, as they take place in aqueous solution where they can take advantage of the solvation energy available for charged or polarized species. Nevertheless, some enzymes do make use of radical reactions, and many of these take place within membranes where less water is available for solvation. For example an alkyl radical is involved in the reactions of vitamin B_{12}. Homolytic breakage of a cobalt–carbon bond (initiation) produces a $-CH_2^{\bullet}$ radical, which subsequently takes a hydrogen atom from another molecule to form a $-CH_3$ group and a second radical (propagation). This is part of a biochemical pathway for the oxidation of fatty acids. For another possible example, see p. 88.

6.7 Radical reactions in the gas phase

Here solvation is not possible and radical reactions often dominate, for example in the atmosphere. The formation of photochemical 'smog' and the formation and destruction of the ozone layer are all complex radical processes.

7 Taking it further

7.1 Introduction

You are now familiar with a good deal of factual functional group chemistry, and when you have worked through the first six chapters of this book you will have met many of the basic ideas that underpin organic chemistry. These ideas help us to relate one aspect of the subject to another. They allow us to thread the facts together, like beads on a necklace, holding them together to make a satisfactory pattern. Such patterns should be recognizable and memorable so that they can be extended and developed in new and interesting ways.

Theories need evidence

Underpinning these patterns is a solid base of fact, that is, *good, repeatable experimental data*. We need theories, but all good theories are firmly based on facts. The mechanisms in Chapters 4 to 6 are *descriptions* of what happens; in very few cases have we given you any of the experimental data upon which they are based or explained the interpretation of the data (but see p. 63). 'How do you know that?' and 'How do you show that this statement is true?' are questions you will be asking more often in the future. It is in answering these questions that the armoury of techniques of the modern organic chemist comes into play, from chemical kinetics to nuclear magnetic resonance (NMR) spectroscopy and X-ray crystallography.

Mechanisms cannot be 'proved'; they are *descriptions* based on the best available data. As new reliable data appear, mechanisms are constantly modified and updated.

7.2 Mechanisms and molecular orbitals

We have considered most reactions in terms of an electrophile ($\delta+$, or 'electron-deficient'), and a nucleophile ($\delta-$, having a nonbonded electron pair or 'electron-rich'), which might imply that the interaction between the two is just the attraction of positive and negative charges. This would be an over-simplification; it is clear that the electron pair concerned must go from an orbital in the nucleophile towards an orbital in the electrophile. The reaction therefore can also be seen as *an interaction between two orbitals*, molecular or atomic, an *occupied* orbital in the nucleophile and an *unoccupied* one in the electrophile. (Remember that each orbital can hold only *two* electrons.) This interaction depends on the size, shape, and energy levels of the two orbitals. If we are dealing with elements in the first short period of the periodic table (Li–F) the size-and-shape match will probably be quite good. Since electrons fill the lowest energy orbitals first, an occupied orbital will normally be of lower energy than an unoccupied one. We are therefore looking for a relatively *high-energy occupied orbital* to interact with a relatively *low-energy unoccupied orbital*.

Nonbonded and π electron pairs occupy relatively high-energy orbitals, and polar (or multiple) bonds such as C—Cl and C=O provide relatively low-energy unoccupied orbitals. So most of our reactions involve the combination of molecules with nonbonded pairs or π electrons, such as NH_3 or ethene, with haloalkanes, carbonyl compounds, or even bromine (permanent or temporary polarization). These are exactly the molecules we have already classified as nucleophiles and electrophiles. So the two ways of looking at these reactions do have common ground!

Orbitals and stereochemistry

Orbitals, which show the probable distribution of electron density, may have directional properties; remember the three mutually perpendicular p orbitals (see p. 2). So for a successful interaction between two orbitals, we also need to have their relative positions in space more or less correct: the reagent has to be on the right 'flight path' to start bond formation. The study of the stereochemistry of interacting molecular orbitals is called *stereo-electronics*. We will look at two examples: first, nucleophilic substitution of a haloalkane (see p. 41).

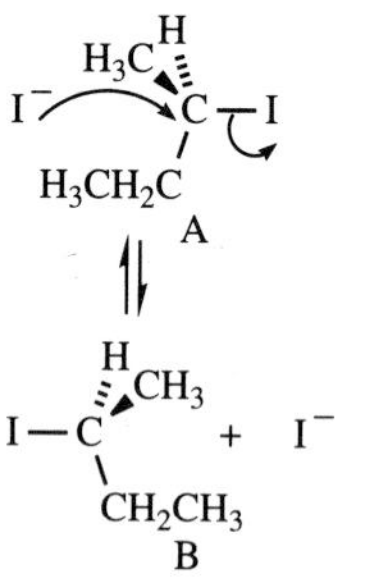

Stereochemistry of nucleophilic substitution. The nucleophile has a nonbonded pair (high-energy filled orbital). The haloalkane has its lowest energy unoccupied orbital at 180° to (that is, behind) the C—Br bond, so the best flight path is for the nucleophile nonbonded pair to approach the back of the C—Br bond.

HO^- ... C—Br ⟶ HO—C + Br^-

This is how they are drawn in Chapter 4 with curly arrows.

HO^- C—Br ⟶ HO—C + Br^-

I^- + H_3C, H, C—I, H_3CH_2C (A) ⇌ H, CH_3, I—C, CH_2CH_3 (B) + I^-

What is the stereochemical relationship between A and B in this substitution reaction? (The energy profile of this reaction looks like the second diagram on p. 24.)

This flight path has important stereochemical consequences if the haloalkane is optically active, because it throws the molecule 'inside out'.

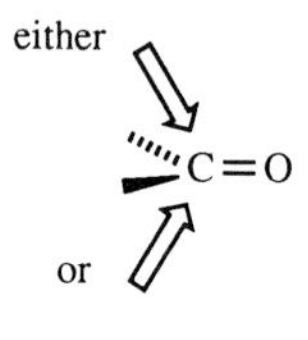

Stereochemistry of addition to carbonyl. As the second example, let's look at addition to the carbonyl group (see p. 47). We believe that nucleophiles attack the carbonyl carbon atom from a direction at right angles to the plane of the carbonyl group and from slightly behind. This is where the lowest-energy unoccupied orbital is located. Some of the experimental evidence for this comes from collecting observations of many different X-ray crystal structure analyses of compounds which include both a carbonyl group and a nucleophile close together, and then looking for the most popular relative positions. These results agree with modern theoretical calculations and with the current molecular orbital picture of the carbonyl group. Now we can use this idea to predict the outcome of a new reaction, or to help us to understand the stereochemistry of other reactions.

either ⇘ C=O ⇗ or

Enzymes can break the symmetry of simple molecules

When 1-deuterioethanal is reduced by $NaBH_4$, an equimolar (1 : 1) mixture of the two optical isomers of 1-deuterioethanol is produced.

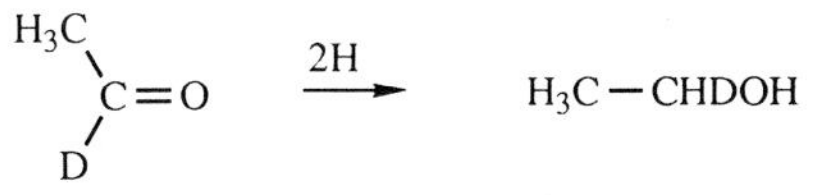

$$H_3C(D)C{=}O \xrightarrow{2H} H_3C{-}CHDOH$$

It is equally likely that the H^- will approach the top or bottom face of the molecule.

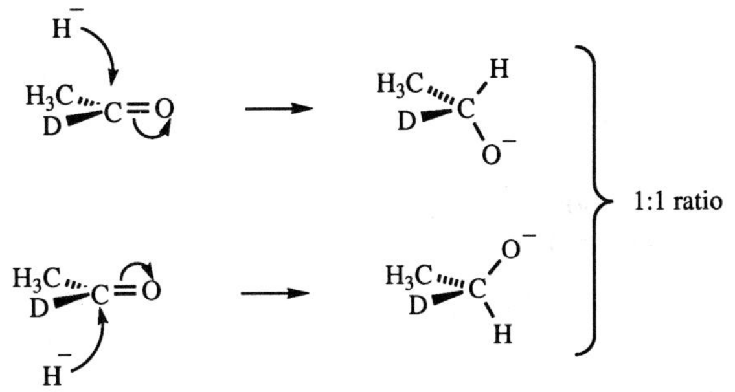

1:1 ratio

Make yourself some molecular models of these molecules to help you to understand the stereochemistry of these reactions.

Two enzymes, liver alcohol dehydrogenase and yeast alcohol dehydrogenase, will also catalyse the reduction of 1-deuterio-ethanal, but there is a striking difference from the sodium borohydride reaction. Each enzymic reaction produces only *one* optical isomer of 1-deuterioethanol.

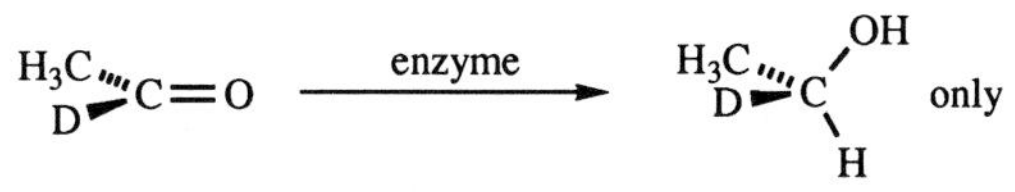

enzyme

only

The deuterioethanal is bound to the enzyme surface, so that the top and bottom faces are not identical; the symmetry has been broken. The reducing H is delivered from a nearby part of the enzyme surface.

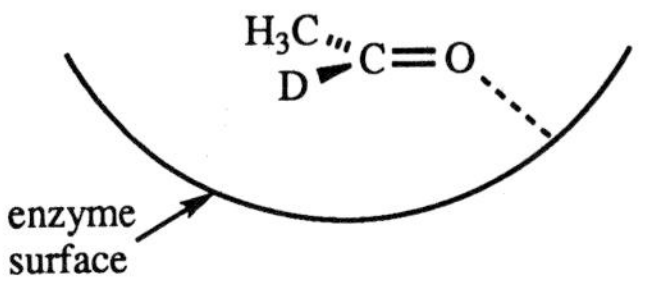

enzyme surface

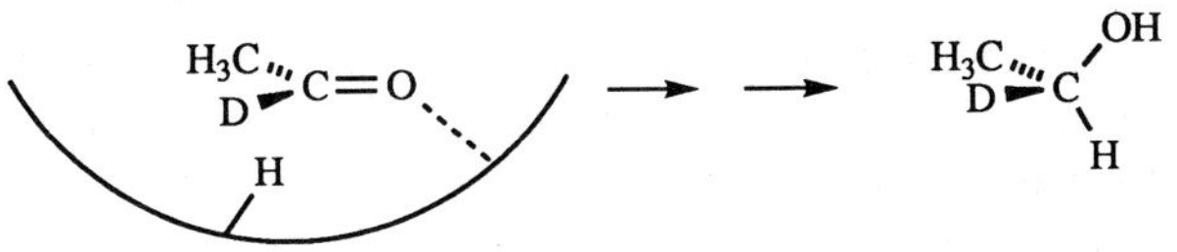

Each enzyme breaks the symmetry of the original planar aldehyde, so that a single, optically active isomer of the alcohol is made. This research gives you an example of the use of isotopes to study the course of a reaction. H and D are sufficiently different that there are two optical isomers of CH_3CHDOH, but sufficiently similar that the chemical reactions of CH_3CHO and CH_3CDO are almost exactly the same.

7.3 Using functional group chemistry

Having looked at the extensions of some theories, now let's look at the application of some familiar reactions on new molecules.

Benzocaine and Novocaine

The local anaesthetic Benzocaine is an ester—an ethyl ester of a substituted benzoic acid.

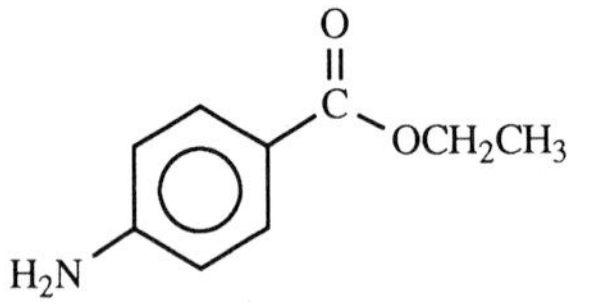

Which of the two nitrogen atoms of Novocaine is the more basic? (See p. 38.)

H_2N–C_6H_4–$C(=O)OCH_2CH_3$

Benzocaine

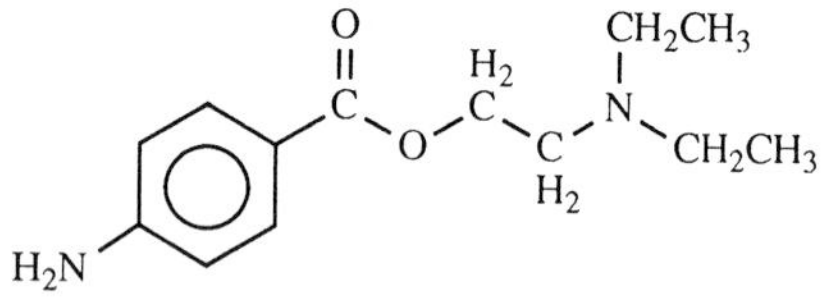

H_2N–C_6H_4–$C(=O)$–O–CH_2–CH_2–$N(CH_2CH_3)_2$

Novocaine

Esters can be made from an acid and an alcohol (see p. 52) so Benzocaine could be prepared from 4-aminobenzoic acid and ethanol. We need an acid catalyst and dry conditions, such as dry HCl. The reactions will be:

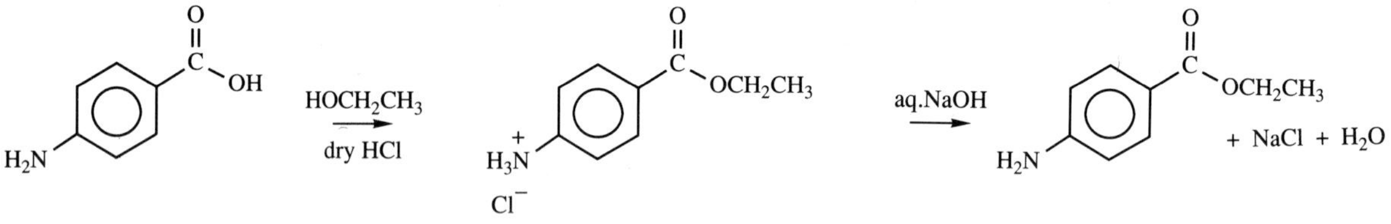

$$H_2N\text{–}C_6H_4\text{–}COOH \xrightarrow[\text{dry HCl}]{HOCH_2CH_3} Cl^-\,\overset{+}{H_3N}\text{–}C_6H_4\text{–}COOCH_2CH_3 \xrightarrow{\text{aq.NaOH}} H_2N\text{–}C_6H_4\text{–}COOCH_2CH_3 + NaCl + H_2O$$

Novocaine is a newer, related local anaesthetic. Work out the last stages for a synthesis of Novocaine.

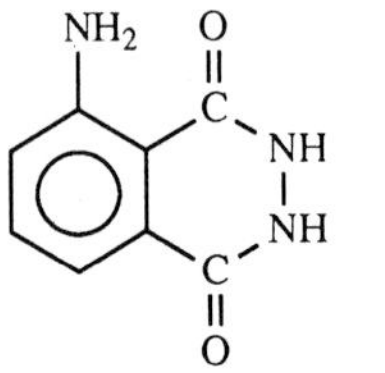

Luminol

NH₂, C=O, NH, NH, C=O (ring structure)

Luminol

Luminol is chemiluminescent. When it decomposes in the presence of alkaline $K_3Fe(III)(CN)_6$ and H_2O_2, it gives out energy in the form of ultraviolet light. The hydrocarbon rubrene can absorb this light energy and re-emit it by fluorescence, making the solution glow with a bright yellow light. This type of reaction can be used in 'light sticks'; the reagents are mixed by breaking the barrier between the two compartments when the stick is bent.

Luminol can be made from benzene-1, 2-dicarboxylic anhydride in three stages.

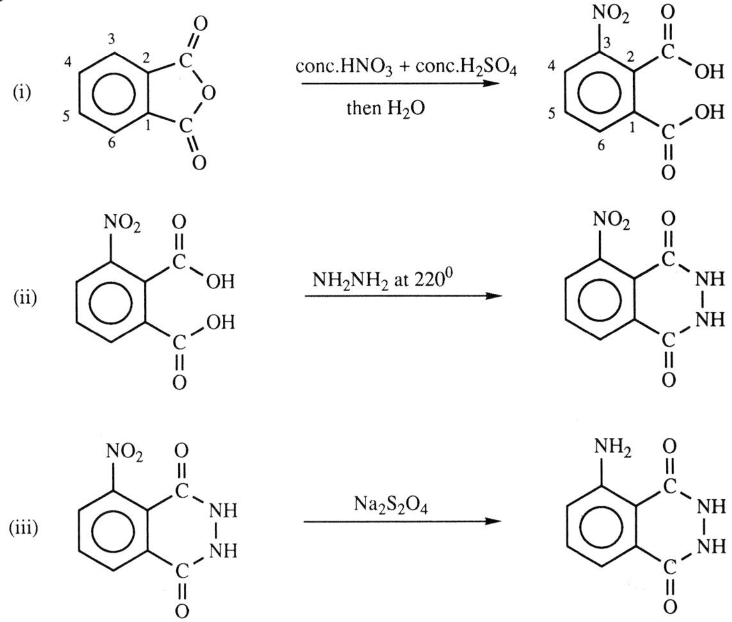

(i) conc.HNO_3 + conc.H_2SO_4 then H_2O

(ii) NH_2NH_2 at 220^0

(iii) $Na_2S_2O_4$

Luminol

Fireflies produce their light by enzymic oxidation of LUCIFERIN.

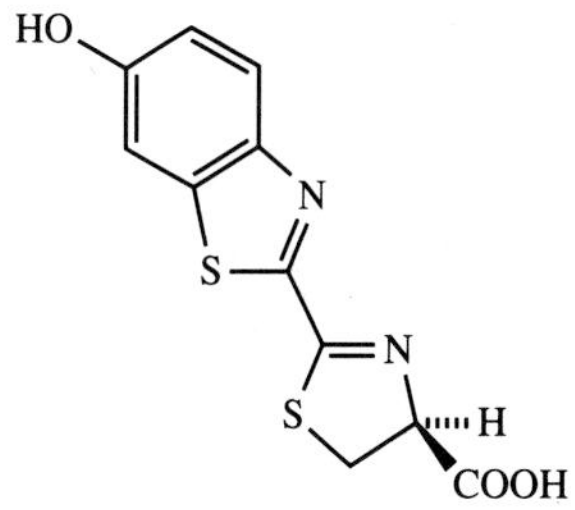

Stage (i) is the nitration of an aromatic compound (see p. 70) followed by the hydrolysis of an acid anhydride (see p. 54); the only difficulty is whether the new nitro group substitutes at position 3 or 4. Draw these mechanisms out.

Stage (ii) is the reaction of a carboxylic acid with an amine under forcing conditions (220°C) to form an amide bond (see p. 55). This happens twice; draw this mechanism, too.

Stage (iii) is the reduction of a nitro group to an amine. The reducing agent $Na_2S_2O_4$ may be new to you; another possible reagent might have been hydrogen with a palladium catalyst, but this breaks the N—N bond as well.

You will notice how we have put each stage into its categories of reaction type (e.g. reduction in stage (iii)) and functional group change (e.g. acid to amide in stage (ii)). This is a very useful way to look at the chemistry of complex molecules, but you learn to be careful of undesirable side-effects on other functional groups (see stage (iii)).

The structure of the whole molecule affects each functional group

We must remember that the rest of the molecule around a functional group affects its chemistry. Our choice of reducing agent in the previous subsection depended on other groups present in the molecule, and we can

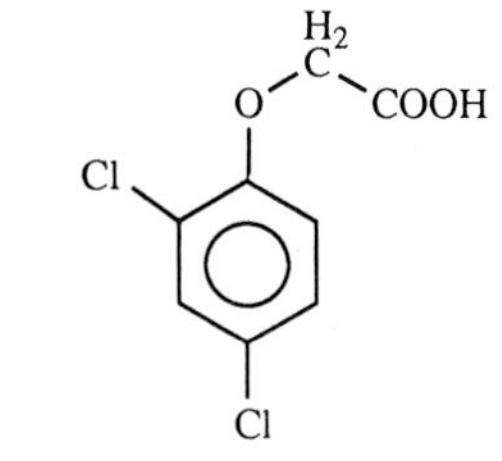

2,4-D

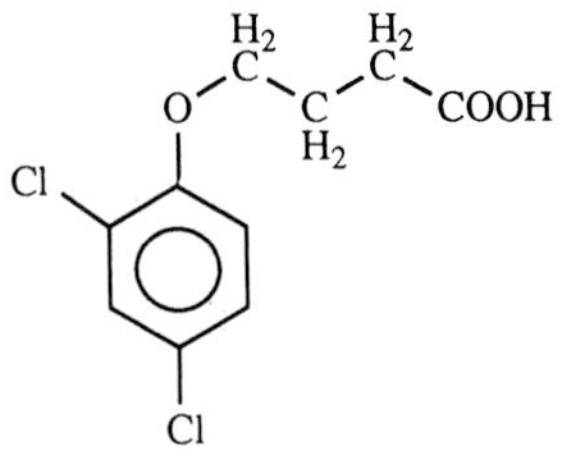

2,4-DB

see how the reactions of bromoethane and of 2-bromo-2-methylpropane with NaOH differ (see p. 46).

$$CH_3CH_2Br + NaOH \longrightarrow CH_3CH_2OH + NaBr$$

$$(CH_3)_3CBr + NaOH \longrightarrow (CH_3)_2C{=}CH_2 + NaBr + H_2O$$

The biochemistry may alter too, even with quite subtle changes. The two molecules 2,4-dichlorophenoxyethanoic acid (2,4-D) and 2,4-dichlorophenoxybutanoic acid (2,4-DB) look very similar chemically but they are very different to a plant.

2,4-D stimulates a fatal level of growth in broadleaved plants and is an excellent weedkiller for narrow-leaved crops like wheat. 2,4-DB itself is inactive as a weedkiller, but some broadleaved plants metabolize it by oxidation to 2,4-D—with fatal effects. Thus 2,4-DB is even more selective than 2,4-D.

7.4 Some current research projects

What sort of things are organic chemists busy investigating now? Three examples follow; we make no apology for the biological slant to these because this reflects the authors' interests!

The biosynthesis of ethene: ripening fruit

If you want to make green tomatoes ripen faster, put a ripe tomato or apple or banana with them. The ripe fruit gives off ethene which catalyses the ripening process. We know how to make ethene from coal or oil or ethanol; but how does a plant make ethene?

Ethene will also accelerate the formation of seed-heads from flowers; so a bunch of carnations will last longer if they are not near a bowl of fruit.

Ethene in plants is made from the cyclic aminoacid A, which itself comes from one of the common 'essential' amino acids, methionine:

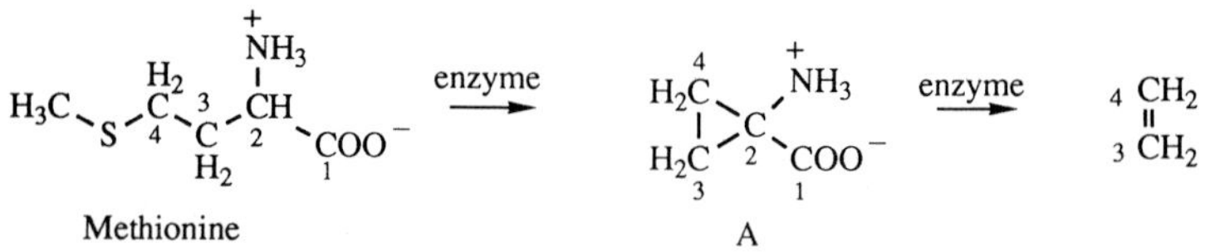

$$H_3C{-}S{-}\overset{4}{C}H_2{-}\overset{3}{C}H_2{-}\overset{2}{C}H(\overset{+}{N}H_3){-}\overset{1}{C}OO^- \xrightarrow{\text{enzyme}} \text{A} \xrightarrow{\text{enzyme}} \overset{4}{C}H_2{=}\overset{3}{C}H_2$$

Methionine

A

How do we know this? If the methionine has been prepared with a radioactive 'label' (^{14}C) at either C_3 or C_4, then this label appears both in the aminoacid A and in the ethene. So the carbon atoms of the ethene are derived from carbons 3 and 4 of methionine. But what is the stereochemistry of the conversion of the amino acid A into ethene? Do the two *cis* hydrogen atoms on the same side of ethene always come from the same side of the ring of A? The answer is *no*—because the two deuterated versions of A both give a 'scrambled' 1:1 mixture of dideuterioethenes (but *no* CH_2CD_2).

The detailed enzyme mechanism is unknown, but it probably involves a radical intermediate. Even less is known about how the ethene catalyses ripening!

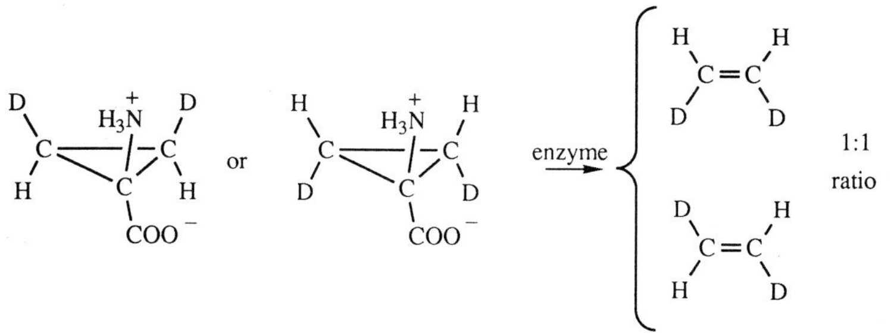

What chemical reaction(s) could you use to differentiate between CH_2CD_2 and CHDCHD? Which of the two geometrical isomers of CHDCHD has the larger dipole moment?

The taxols: synthesis of new anti-cancer agents

One of the most promising new drugs for the treatment of cancer of the breast or colon is taxol. Taxol is extracted from the Pacific yew, and it takes one mature tree to provide enough taxol for one patient. The tree is a slow-growing North American evergreen which is becoming rare. Total synthesis of taxol is difficult because of its complex stereochemistry. Several companies are attempting to set up commercial tissue cultures of Pacific yew bark cells to make taxol. A more wide-ranging approach to the problem is to use a tissue-culture to produce compounds similar to taxol, in which the plant cells have done much of the difficult chemistry. For example baccatin IV is *comparatively* abundant in English yew, and other related species have yet to be scanned.

Make a list of the names and structures of all the functional groups in Taxol.

Taxol

(0.01% in bark)

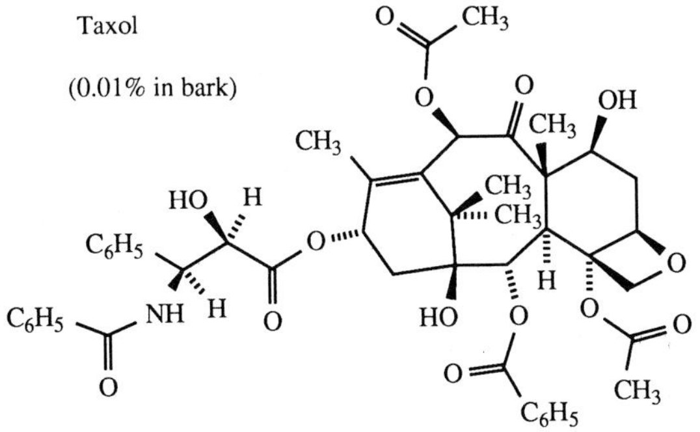

Baccatin IV

(1% in leaf)

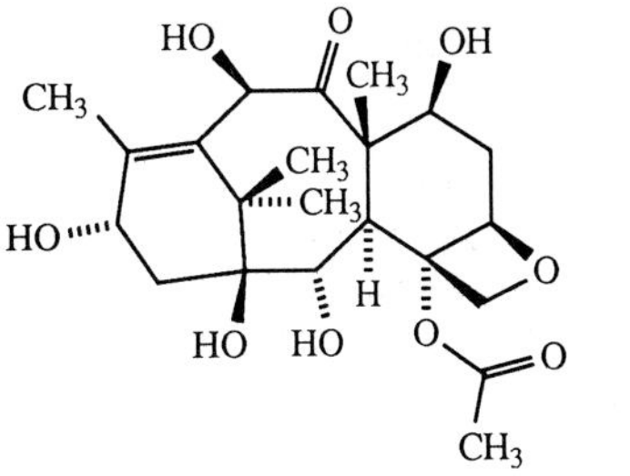

These and similar compounds can then be elaborated in the laboratory to give a new range of potential anti-cancer drugs. The powerful combination of cell culture and organic synthesis is a flexible approach which may help to save the lives both of the patients and of the Pacific yew.

What reactions would be needed to convert Baccatin IV into Taxol? List them as functional group inter-conversions, e.g.
secondary alcohol $\xrightarrow{\text{oxidation}}$ ketone.

Tools for biochemistry and physiology: fluorescent indicators for calcium ions

How does a heart beat? How does the beating of a diseased heart differ from that of a healthy heart, and why? What is the molecular mechanism of the contraction of a single heart cell? Changing the concentration of calcium ions in the cell is one of the control mechanisms, and so a method is needed to measure this inside the cell.

Methyl orange and other pH (proton) indicators show different colours when they are bound to protons (in acid solution) and when they are not (in alkaline solution).

Indicator (alkaline solution) ORANGE	⇌	Indicator H^+ (acidic solution) RED

Calcium indicators will do the same with calcium ions, Ca^{2+}. The most sensitive indicators are fluorescent and show a different fluorescence colour depending on whether they are bound to Ca^{2+} or not. These indicators are complex molecules with two overlapping parts: one to bind the Ca^{2+} and one to give the fluorescence, e.g. INDO–1.

binding

5 K^+

INDO-1

fluorescence

INDO is a large organic molecule, and yet it is soluble in water: why? And to which atoms in INDO do you think the calcium ion binds?

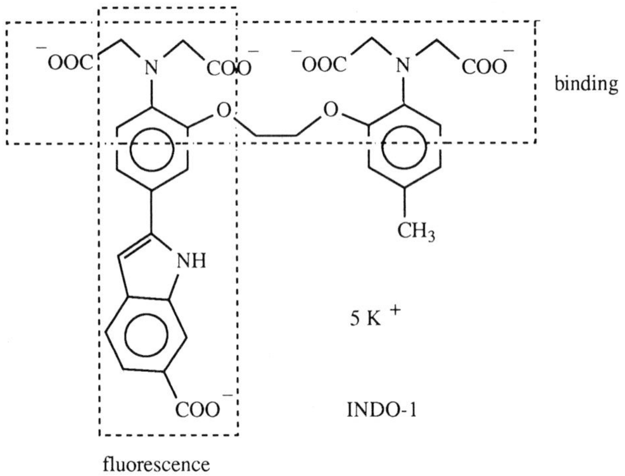

Calcium ions are very important in cells such as nerve cells and in the retina of the eye. Calcium ions are also involved in the opening and closing of the stomata in leaves.

New and better indicators are being synthesized for Ca^{2+} as well as for other biologically important ions such as Mg^{2+} and Zn^{2+}, which will be very valuable for medical research. The collaboration between the synthetic organic chemists and the physiologists is crucial for progress.

7.5 Conclusion

Almost every section in this book could have been developed further, with more sophisticated ideas and examples. From time to time you will have glimpsed the depths and sensed their challenge, a challenge which we hope you now feel ready to take up.

Further Reading

General A level chemistry: Atkins, P. W., Clugston, M. J., Frazer, M. J., and Jones, R. A. Y. (1988). *Chemistry: principles and applications*. Longman, London.

Undergraduate organic chemistry: Kemp, D. S. and Vellaccio, F. (1980). *Organic Chemistry*. Worth, New York.

Biological Aspects: Stryer, L. (1988). *Biochemistry*. 3rd edn, Freeman, New York.

Research topics in Chapter 7:

(a) Ethene: Adlington, R. M., Baldwin, J. E., and Rawlings, B. J. (1983). *J. Chem. Soc. Chem. Comm.*, 290.

(b) Taxol: Potier, P. (1992). *Chem. Soc. Rev.* **21**, 113.

(c) INDO: Grynkiewicz, G., Poenie, M., and Tsien, R. Y. (1985). *J. Biol. Chem.* **260**, 3440.

Index

acid
  anhydrides with nucleophiles 54
  chlorides, *see* acyl chlorides
  conjugate 34
  dissociation 33
  strength 33–8
  strong/weak 33
activation energy 24–5
acyl chlorides with nucleophiles 51, 54–5
addition 18, 58
addition–elimination 49–50, 56
alcohols
  elimination 18
  hydrogen bonding in 16, 17
  as nucleophiles 54
  nucleophilic substitution 44
  oxidation 31
aldehydes
  addition–elimination 49–50
  enzymic reduction/stereochemistry 84–5
  nucleophilic addition 47–8
  reduction 47–8
  structure 4
alkanes
  cracking 79
  radical substitution 75–8
  solubility 17
alkenes
  electrophilic addition 58–66
  hydrogenation 78
  oxidation 66
  polymerization 64–5, 79–81
  structure 4
alkyne structure 5
amide hydrolysis 53–4
amines
  as nucleophiles 43–4, 54, 55
  as weak bases 34, 38
amino acid structure 38–9
Arrhenius equation 27
asymmetric carbon 14

bases 33
  as leaving groups 35–6
  as nucleophiles 35–6, 40
  strength 38
benzene
  compared to alkenes 74
  electrophilic substitution 69–74
  reduction 68
  structure 5, 67–9
boiling points and structures 15–16
bond
  angles 11–12
  lengths and strengths 10–11
bromination 62, 71, 77
bromonium ion 62
Brønsted–Lowry theory 33

CFCs 11
carbocations 12, 29, 45–6, 58
  stability 60–1
carbonyl compounds, *see* aldehydes *or* ketones
carboxylic acids
  addition/elimination 55
  esterification 53
  reduction 55
  solubility 16–17
  as weak acids 33
  see also acid
catalysis 26, 29
chain reactions, *see* radical reactions
chirality 14
chlorination 64, 72, 75
cis/*trans* isomerism 13–14
collision theory 24–5
condensation, *see* addition–elimination
conjugated double bonds 6
cracking 79
curly arrows 20–2

dative covalency 10, 59
delocalization 5, 37–8, 62, 67–9
diazotization 73
dipole–dipole attraction 15
dipoles 6, 16–17
dot diagrams 9
dynamic equilibrium 23

electronegativity 6, 37, 60
electrophiles 19, 58
electrophilic
  addition 58–66
  substitution 69–74
elimination 18, 46–7
energy
  barrier (activation energy) 25, 28
  changes ($RT\ln K$) 23
  profiles 23–5, 28–9
equilibria 22–3, 33–5
esterification 52–3, 86
ester
  hydrolysis 52
  reduction 53

fishhooks 20–1
functional groups 8–9, 86–7

geometric isomerism 13–14

haloalkanes
  elimination 32, 46–7
  nucleophilic substitution 32, 41–7
    stereochemistry of 84
  polarity 7
halocompounds, comparative
    reactivity 56–7
heterolysis 18, 41, 59
homolysis 18, 75
hybrid orbitals 3
hydration 61
hydride ion donors ($NaBH_4$, $LiAlH_4$) 47,
    53
hydrogen bonding 16–17
hydrogenation 78–9
hydrolysis 43

inductive effect 7, 37, 60–1
intermediates 9, 28–9, 45–6, 47
intermolecular attractions 15–17
ion–dipole attraction 15
isomerism 12–14

K_a and pK_a 33, 34
K_b and pK_b 34
ketones
  addition–elimination 49–50
  intermolecular attraction 15
  nucleophilic addition 47–8
  polarity 15
  reduction 47–8
  solubility 17
  structure 4
kinetic
  control 29–31
  energy of molecules 25–6
kinetics 26–31
K_w 34

leaving groups 35–6, 50
lone pairs, *see* nonbonded pairs

Markovnikov's rule 61
Maxwell–Boltzmann diagram 26
molecular orbitals 2, 83
multiple bonds 3–5

nitration 70, 87
nitriles
  hydrolysis 36, 43, 48
  reduction 43, 48
nonbonded pairs 10
nucleophiles 10, 19, 40
nucleophilic
  addition 47–8
  addition–elimination 49–55
  reactivity 35–6
  substitution 41–7

optical isomerism 14
order of reaction 27
oxidation of
  alcohols 31
  aldehydes 31
  alkenes 66
  benzene compounds 74
ozonolysis 66

p orbitals 2
pH 34
  and amino acid structure 38–9
partial charges ($\delta+$ and $\delta-$) 6
phenol
  electrophilic substitution 72, 73
  as a weak acid 37
phenylamine
  diazotization 73
  electrophilic substitution 72
  as a weak base 38
photochemical reactions 75–8
pi (π) bonds 4
polarization of bonds 6, 15, 19, 40
polymerization 55, 64–5, 79
  cationic 64–5
  radical 79–81
proton (H^+) 33, 61
protonation 44, 47

radical 18
  flow diagram 77
  reactions 22, 75–82
  stability 80
rate
  constant 27
  determining step 27–8
  equations 26–8
reduction of
  acids 55
  aldehydes 47
  alkenes and benzene 78–9
  esters 53
  nitriles 43, 48
  nitrocompounds 71, 87
resonance 5, 22, 67

s orbitals 2
shapes of molecules 3, 4, 11–12
  drawing in 3D 11
sigma (σ) bonds 2
solubility
  of acids and bases 35
  factors affecting 16–17
solvation 37
stereoelectronics 84
stereoisomerism 11–14
structural
  formulae 7
  isomers 12
substitution 18, 41, 69, 75

thermodynamic control 29–31
transition states TS 12, 28–9, 42

van der Waals forces 15–16